ALFRED WRIGHT'S AUTOGRAPH BOOK

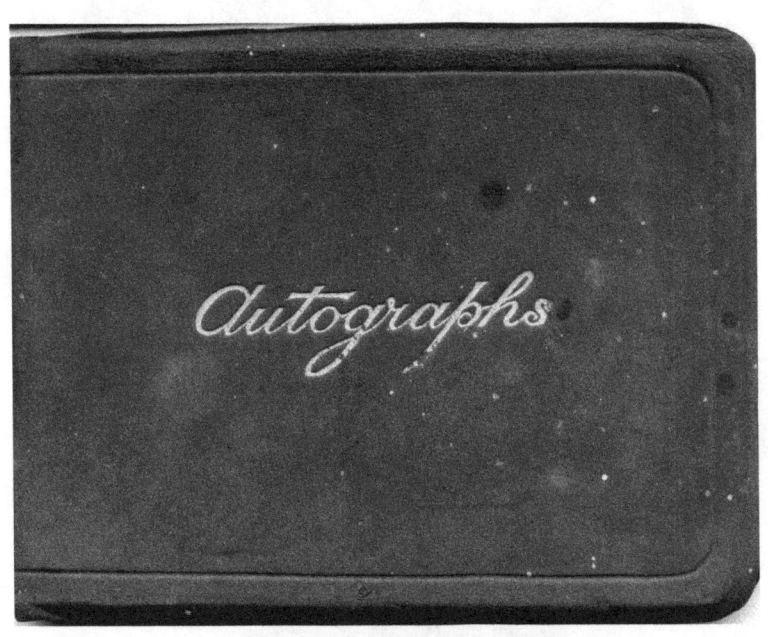

Alfred Richmond Wright.

ALFRED WRIGHT'S AUTOGRAPH BOOK

Chris Edye

ETT IMPRINT
ExileBay

First published by ETT Imprint, Exile Bay 2020

Copyright © Chris Edye 2020

This book is copyright. Apart from any fair dealing for the purposes of private study, research criticism or review, as permitted under the Copyright Act, no part may be reproduced by any process without written permission. Inquiries should be addressed to the publishers:

ETT IMPRINT
PO Box R1906
Royal Exchange NSW 1225
Australia

ISBN 978-1-922473-11-0 (paper)
ISBN 978-1-922473-12-7 (ebook)

ISBN 978-1-922473-40-0 (trade)

Cover design by Tom Thompson

CONTENTS

1 Introduction	6
2 Preface	7
3 Two Special Pages from the Album	9
4 Alfred's Early Life	10
5 Alfred Enlists	11
6 Alfred Goes to War	16
7 Signatures on the Voyage	20
8 Back on Dry Land	25
9 Wright in France	26
10 After the War	42
11 Alfred's Death	47
12 Extended Biographical Notes	49
13 The State Aviation School at Richmond	76
14 Picture Credits	78
15 A little bit of fun at the end	79
16 End notes	80
17 Index	83

1
Introduction

Alfred Richmond Wright was a 2nd Lieutenant travelling aboard the *Euripides* from Australia to England to the Great War in 1917 when he bought an autograph book. Leather-covered, gold-edged, with 64 pages, it probably wasn't cheap when Alfred bought it, but now, over 100 years later, it is priceless. It is in remarkably good condition, given its age and its history, particularly its early history, which is a tribute to the care that it has received over the years.

Alfred bought the autograph book either on or just before 7 December 1917. As you do when you've got something new, you show it to your friends and colleagues, and, it being an autograph book, they sign it. With any luck, they also write when and where they signed it.

This is the story of Alfred's progress through his war, and of the people he met who signed his book.

2
Preface

There are 133 signatures in Alfred's autograph book. Almost all have further details of the signatory, such as their rank and unit, their location at the time they signed, and their home address.
The important characteristic of a signature is that it is unique, not that it is legible, and many of those who signed the book clearly embraced this principle. As a result, a certain amount of detective work has been required to identify each signatory and transcribe the information they provided. Success in this process has not been complete.

The information used to provide details of each signatory has come from a range of sources, some civilian but mostly military. These include – the records of service held by the National Archives of Australia; the embarkation and nominal rolls held by the Australian War Memorial; unit war diaries and photographs held by the AWM; Honour citations and awards held by the AWM. In addition, the Commonwealth and NSW Government Gazettes, the NSW Births Deaths and Marriages on-line facility, and sundry newspaper reports accessed through the National Library of Australia's Trove facility have provided essential and background information.

I have used endnotes extensively, not footnotes because of the interruption they cause to the flow of reading, and because of the family nature of this document. All the information presented, however, is drawn from the sources noted above, and can be found there should the reader wish. I have referred to the subject of this piece as 'Alfred' in the family context, and after the war, and 'Wright' during the war section, to make it consistent with referring to his war colleagues by their surnames.

The task would not have been possible without the generosity of Alfred's granddaughter Val and her husband Ian Thom, who lent me the book and Alfred's other records and photographs. They allowed me to scan them and provided encouragement and further information about him. Only where a photograph is drawn from elsewhere is this indicated.

Abbreviations

AAMC	Australian Army Medical Corps
AAVC	Australian Army Veterinary Corps
AE	Australian Engineers
AIF	Australian Imperial Force
AMC	Army Medical Corps [Australia]
CB	Companion of the Bath
CBE	Commander of the Most Excellent Order of the British Empire
CdeG	Croix de Guerre [French honour awarded by the President of France]
CMF	Citizen Military Forces
CMG	Companion of the Most Distinguished Order of St Michael and St George
DSO	Distinguished Service Order
EOTS	Engineer Officer Training School, Roseville NSW
FCE	Field Company Engineers
HMAT	His Majesty's Australian Transport [ship]
HT	Hospital Transport [ship]
MC	Military Cross
MM	Military Medal
OBE	Officer of the Most Excellent Order of the British Empire
RFC	Royal Flying Corps
ROS	Record of Service
VD	Volunteer Officers' Decoration
YMCA	Young Men's Christian Association

3

Two Special Pages from the Album

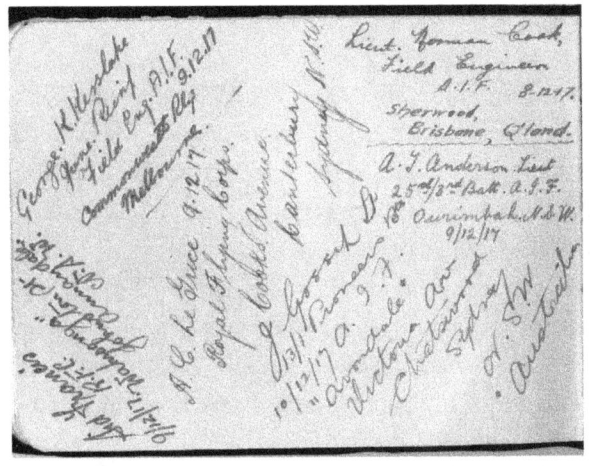

Album opening (20).

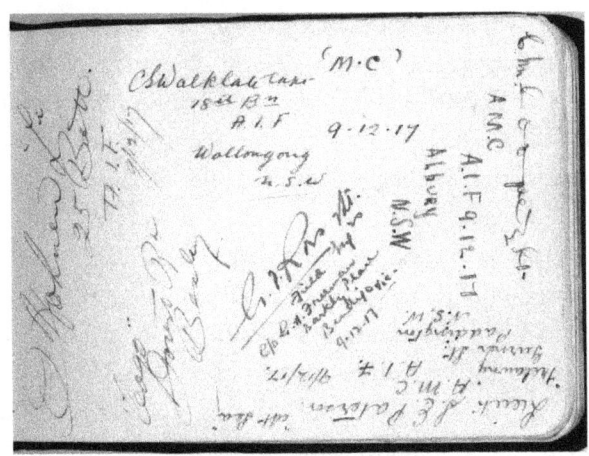

Album opening (16).

4

Alfred's Early Life

Alfred Richmond Wright was born on 15 September 1889 to Alfred and Clara Wright (nee Carr) in Lismore, NSW. He was the only son and second child, his sisters being Clara (b. 1886 in Balmain), Amy (b. 1893 in Lismore) and Kathleen (b. 1902 in St Peters). He spent much of his boyhood in Ballina.[1]

His second given name is the name of the river that runs through Casino and empties into the South Pacific Ocean at Ballina. Apart from this, Alfred retained a link to the area of his birth throughout his life - his parents and sisters lived in the area, and he later did business there.

Alfred studied at Fort Street High School in Sydney. At the time he enlisted on 10 May 1916 he gave his trade or calling on his attestation form as 'Constructional Engineer' but without any indication as to where he undertook the training for this. He also noted that he was still serving as a member of 1 Regiment, 9 Brigade CMF [Citizen Military Forces]. Later, on his application for a commission in the AIF on 12 December 1916, he was more forthcoming: he had spent 4 ½ years in 1 Regiment Scottish Rifles and was a Lieutenant in 36 Battalion 9 Brigade CMF.

In 1910, in Ballina, Alfred married Charlotte Agnes Valentin. They had three children – Dorothea Daphne (b. 1910 in Ballina), Valerie (b. 1913 in Petersham), and Rex (b. 1916 in Burwood). On 7 September 1911 Alfred was appointed a police officer in NSW.[2]

On 1 January 1916 Alfred was appointed a Lieutenant (provisional) in the Commonwealth Military Cadet Corps, 2 Military District [NSW].[3]

5

Alfred Enlists

On 10 May 1916 Alfred enlisted in the AIF and was immediately assigned to the Engineer Officer Training School at Moore Park, in Sydney's Eastern suburbs. He attended a series of lectures there given by Lieutenant Morton, two of which were delivered before Alfred had actually joined up. The lecture notes he wrote up begin with a lecture on 'training of an officer' on 8 May and end with 'night operations' on 1 June.

At about that time the EOTS moved to Roseville, on Sydney's north shore, and Alfred moved with it. Details of his time at Roseville are incomplete, not to say non-existent: Alfred's 'Record of service in the field' and his 'Casualty Form – Active Service,' documents completed by the AIF to show where a soldier was at any particular time, make no mention of this period.

Fortunately, family records make up for this lack, and photographs Alfred took with his vest pocket autographic Kodak camera (below; still in the family's possession) show a broader picture of life at EOTS Roseville.

Honest Toilers.
Smoko Sport.
Two photographs taken on September 14 1917.

On 12 December 1916 Alfred lodged an application for a commission in the AIF in which he noted his previous military history, and that he had spent seven months at the Engineer Officer Training School Roseville. This is the first, and only, recognition of the time he spent there.

Alfred was appointed 2nd Lieutenant on 15 January 1917. His record is similarly incomplete in recording what he did between then and the end of October, but the photograph of No. 1 section at EOTS Roseville (page15) gives a clue: he was on the staff there. The official caption to the photograph, which is not shown in this rendering, lists him as 'Lieutenant', which would have been premature, since the photograph was dated 30 October 1916, but the rank would have given him the authority to instruct students at the school.

2nd Lieutenant Wright with his kit.

Photographs in the family collection show some of the activities undertaken by the students and staff at the school. Whilst evidence from other sources indicates that the men at the School worked hard, the family collection photographs indicate that there was a good spirit and a sense of camaraderie, the adoption of a unit motto and the creation and preservation of a unit flag being good examples.

"Time for a Cuppa."

Soldier's mess at Roseville Engineer Officer Training School.

(Top) Engineer Officer Training School No. 1 Section at Roseville, NSW 30 October 1916. Alfred is standing on the right. Captain John Madsen is seated in the middle of the second row. (Lower) Bridge building was a big part of the work done at Roseville. There was a saying that 'an engineer never saw a river he didn't want to bridge, or a bridge he didn't want to demolish,' and this seems to have been well understood by the students.

6

Alfred Goes to War

On 31 October 1917 Alfred embarked aboard HMAT A14 *Euripides* for the voyage to England as the senior 2nd Lieutenant in charge of the June 1917 reinforcements for the Field Companies Engineers, with 2nd Lieutenants George Percy Ross and George Knight Kerslake. Also on board were the July 1917 reinforcements for the Field Companies Engineers, with 2nd Lieutenants Edward Charles White, Norman Cook and Eric Houghton Rhodes. There were 270 officers and other ranks in total in the Engineer reinforcements.

The EOTS graduation group photo of the June 1917 reinforcements. Alfred is in the third row, behind the bass drum and 2nd from the left of the fold, with Madsen to his right.

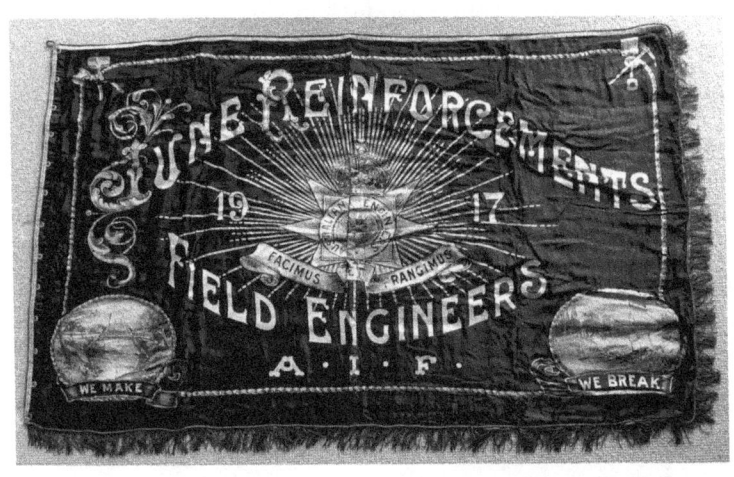

Field Engineers June Reinforcements 1917 Unit flag. The motto
'Facimus, Frangimus' translates as 'We make, we break'
HMAT A14 *Euripides*

The *Euripides* could accommodate 136 officers and 2204 other ranks in its troopship configuration. On this voyage there were 1341 officers and other ranks in reinforcements for several infantry battalions, a Pioneer battalion, 2 divisional signal companies and the Army medical corps, as well as chaplains, dental details, medical officers and sea transport staff; in total, according to the embarkation lists, there were 1876 aboard.

The voyage was long, and a major challenge was to keep the men from becoming bored. Training and fitness work became particularly important. So what did the engineers do? They built a model bridge. Family pictures show that it was no small accomplishment.

The men who built the bridge show it off to the men from the bridge.

The voyage of the Euripides

The *Euripides* departed Sydney on 31 October 1917, heading east. Herbert Austin Bailey, a sapper with the 15th Field Company, Australian Engineers, kept a diary of the voyage. He related that the ship had a relatively smooth journey from Sydney towards Panama, where it transitted the canal and stopped at Colon to take on coal. While the ship was there another ship from Australia, the HMAT A60 *Aeneas*, which had departed Melbourne on 30 October, came into harbour and took on coal when *Euripides* had finished doing so. Both ships then sailed for Port of Spain, in Trinidad, the *Euripides* reaching there on 2 December. After some delay caused by the weather, the men aboard the *Euripides* were allowed ashore on 5 December, where they received a very warm welcome, and set about ridding themselves of cabin fever. After two days, however, things became a little ragged, and the men were confined to the ship. On 9 December, the *Euripides*, the *Aeneas* and a French cruiser sent to escort them left for the open sea. Off Grenada they linked up with the *Magderlene* (sic), a steamer taking troops from Trinidad and Grenada to England, and a day later, were joined by a cargo and passenger ship, the *Perou* (sic), to make a small convoy of 5 ships to cross the Atlantic. On 20 December the convoy passed what was thought to be the Azores, although this identification is not definitive. (Diary of Herbert Austin Bailey, at https://www.awm.gov.au/collection/C2620337 accessed 20200207, cover of diary below)

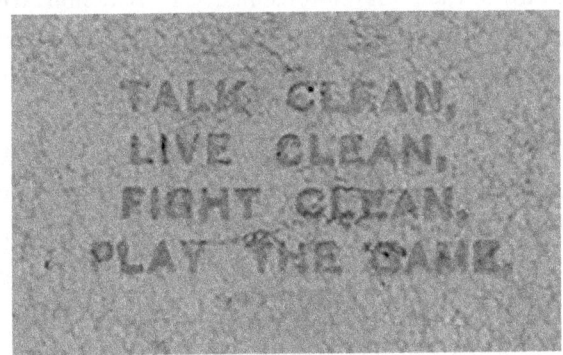

7

Signatures on the Voyage

7 December

The first person to sign the autograph book was Gladys I Stone (see opening 2 of the book), of "Hillaby", on the Long Circular [Road], Port of Spain, on 7 December.

8 December

On 8 December, H Keunga Stone[4] (2)[5], of the same address, also signed it, including the wish 'Good luck!!!'. It is not known who the Stones were, but it is unlikely that the officers aboard the *Euripides* would have been taken to meet people of little importance in Port of Spain when the ship stopped for re-provisioning and re-fuelling; on this logic the Stones were probably senior colonial officials, especially since their address was in as desirable an area in Port of Spain then as it is now.

Second Lieutenant Norman Cook[6] (20) of the July reinforcements of the Field Companies Engineers and 2nd Lieutenant George Henry Smith (3) of the 26th reinforcements, 15 Battalion AIF, also signed the book on 8 December. Cook and Smith had embarked aboard the *Euripides* on the same day as Wright, with their respective units.

Though his signature is undated, this is the only opportunity that Captain Adolph Rudolph William Buttner[7] (22) and Wright would have been in the same location to enable Buttner to sign the book. The same applies to E G Melville (28), who gave his address as Trinidad, British West Indies.

9 December
George Percy Ross (16) and **George Knight Kerslake** (20), Wright's fellow 2nd Lieutenants with the June 1917 reinforcements, both signed it on 9 December. George Ross and Wright had a special bond: their promotions to 2nd Lieutenant had been gazetted on the same date, 8 February 1917, to have effect from 15 January 1917.[8] George was also a witness to Wright's signing of his will.[9] Second Lieutenants **Frederick Henry Hohnen** (16), Smith's colleague with the 26th reinforcements, 15 Battalion AIF, and **Eric Houghton Rhodes** (11), Cook's colleague with the July reinforcements of the Field Companies Engineers, also signed it on 9 December.

The other signatories that day were –
2nd Lieutenant Alick Tonkin Anderson (20) 25th reinforcements, 3 Battalion AIF, a farmer from Ourimbah NSW
2nd Lieutenant Herbert Floyd Bailey (6) 25th reinforcements, 3 Battalion AIF, a planter from Fiji
Captain Samuel Boake (28) Australian Army Medical Corps, a medical practitioner from Mudgee NSW
2nd Lieutenant John Vernon Cobcroft (10) 10th reinforcements, 54 Battalion AIF, a rubber planter from Samoa
Lieutenant Cyril Mansfield Cooper (16) AMC, a dentist from Darlinghurst NSW
2nd Lieutenant Agustus John Cowled (10) 25th reinforcements, 4 Battalion AIF, a farmer from Haberfield NSW
2nd Lieutenant George Percival Darlow (11) 10th reinforcements, 56 Battalion AIF, an indent agent from Randwick NSW
2nd Lieutenant Frederick William Fifield (11) 21st reinforcements, 17 Battalion AIF, a schoolteacher from Narrandera NSW
2nd Lieutenant James Heatley Finlay (10) 26th reinforcements, 9 Battalion AIF, a traveller from Toowoomba, Queensland, who had previously served with 9 Battalion
Sid Francis (20) RFC
2nd Lieutenant Robert Baynton Kerr (10) 10th reinforcements, 55 Battalion AIF, a clerk from Darling Point NSW
A C Le Grice (20) RFC
Captain Jonathan M Maclean (11) Australian Army Medical Corps, a

medical practitioner from Chatswood NSW
Lieutenant S E Paterson (16) AMC, a dentist from Paddington, NSW
Lieutenant (Honorary Captain) William Gould Penrose (5) 26th reinforcements, 9 Battalion AIF, a landowner from Grafton NSW
2nd Lieutenant William Saunderson (11) 10th reinforcements, 56 Battalion AIF, a schoolteacher from Summer Hill NSW
Captain Chaplain Grosvenor Francis Stopford (11) Church of England minister from Maclean NSW
Lieutenant W H Taylor (28) The Queens Royal West Surrey Imperial Army Lieutenant Stanley George Viccars (7) 10th reinforcements, 54 Battalion AIF, a station overseer of Sydney
Lieutenant Harrild Martin Walker (5) 10th reinforcements, 52 Battalion AIF, an insurance manager from Brisbane Queensland
Captain Charles George Walklate, MC (16) 25th reinforcements, 4 Battalion AIF, a crane driver from Wollongong who had previously served with the AN&MEF in New Guinea

10th December
Edward Charles White (25), colleague to Cook and Rhodes, signed the book on 10 December. Also signing that day were –
Lew Cole (7) RFC
Lieutenant William Montgomerie Fleming (3) Lieutenant for the voyage only, eventually became a driver with the Australian Army Service Corps; grazier, of St Kilda, Victoria
2nd Lieutenant James Gossip (20) 13th reinforcements,
1 Pioneer Battalion AIF, a licensed surveyor from Chatswood NSW
Garnet Halloran (23) No military association on the voyage, a medical practitioner from Mosman NSW who had served in France earlier in the war
Lieutenant William David Harris, MC (6) 54 Battalion AIF, a wool traveller and auctioneer of Mosman, NSW
Captain Alfred Joseph Hope (23) AMC, a medical practitioner from Darlinghurst NSW
S E Iggledon (25) C/Officer of the *Euripides*

Lieutenant Joseph Lecky (25) 21st reinforcements, 17 Battalion, a farmer from Homebush NSW
Brian Lucy (7) RFC
Captain John Thomas Moran, MC (23) Soldier, not attached to a particular unit but assigned to 43 Battalion on arrival at Devonport: from Rose Park, SA
Roger Nickoll (7) RFC (the son of a doctor at Mudgee who would have known Samuel Boake as a professional colleague)
Burton Sampson (7) RFC
Captain Lance Hayward Stanton-Cook (23)[10] AMC, a medical practitioner from Turramurra NSW
2nd Lieutenant William Junius Thomas (18) 10th reinforcements, 52 Battalion AIF, a tailor's cutter from Enoggera Queensland
2nd Lieutenant Edward Wauhope (18) 13th reinforcements, 1 Pioneer Battalion AIF, an engineering draughtsman from Randwick NSW.

Several of those who signed in the first few days of December – Lucy, for example - indicated that they were from the Royal Flying Corps, without specifying a rank. The truth is different: they weren't from the Royal Flying Corps at all, well, not yet, anyhow. These twelve were recent graduates from the Aviation School established by the NSW State Government at Richmond, NSW. The story of the Aviation School is set out below.

11 - 14 December
G Dobell (36), RFC, was the only person to sign the book on 11 December 1917. He also wrote 'Atlantic Ocean' which, whilst not being very specific, does tell us that the ship had left Port of Spain. It was in the Sargasso Sea, the mid-Atlantic place to which eels from North America and Europe return to spawn, that Lieutenant **Francis William Broughton** (36) of the AMC signed the book on 14 December.

17 December
By 17 December the ship was off the Azores, which is where Cyril Worboys (36), one of the RFC group, signed it. That day Messrs A Le Sapper (36), 4th officer of the yacht Yolande, and J C Stewart Smith (36), 3rd officer of the *Euripides*, also signed. M A Watts (12), another of the RFC group, signed the book 'off the coast of Spain' but did not show the date of his signature.

23 December
On 23 December the ship was in the Bay of Biscay, to the west of France, and here Alan Edwards (12) and Raynes Royle (36)[11], both from the RFC group, signed. Augustus Bargwanna (12), another of the RFC group, signed in the Bay of Biscay, probably about this time. Mr T Thorburn, a representative of the YMCA, also signed on 23 December.

Some of the signers aboard the *Euripides* did not indicate where or when they signed:

Captain **David Christie** (25)
AAMC, medical practitioner from Narrandera, NSW
Major William Rudolph Clay (22)
AAMC No. 2 Section, Sea Transport Staff, a medical practitioner of Manly, NSW
Colonel Charles Alfred Jenkins, Chaplain (22)
Ship's Chaplain (at least for the Methodists; there were other Chaplains for the Roman Catholics and the Anglicans), Methodist minister of Claremont, WA
Lieutenant Dispenser Alfred Joseph Newton (25)
No. 2 Section, Sea Transport Staff, a pharmacist of Katoomba, NSW
Captain Sidney Albert Pinkstone (6)
55 Battalion, a journalist from Cootamundra, NSW
Lieutenant Colonel Stephen Richard Harricks Roberts, DSO VD (22)
11 Battalion, a civil servant of Subiaco, WA
Keith Williamson (22)
Commander of the *Euripides*

8

Back on Dry Land

Once he had disembarked from the *Euripides* on 26 December, Wright's next posting was to Parkhouse no. 3 Camp on Salisbury Plain the same day. Here he met -

Captain **Gerald Mosman Carr,** OBE (27)
24 Battalion, architect of Sydney, 7 January
Lieutenant **George Hilfers Koch** (27)
9 Battalion, clerk, of Brisbane, 18 January 1918 (the day he returned to his battalion in France)
Lieutenant **Alexander Duncan McArthur** (4)
10 Training Battalion, nurseryman of St Kilda, 15 January Lieutenant **Ernest Vincent Raymont** (12)
3 Pioneer Battalion, farmer of Woody Point, Brisbane, 11 January 2nd Lieutenant **Frank Meredith Turton** (15)
General infantry reinforcements, clerk of Cannington, WA, 18 January 1918

On 30 January 1918 Wright was admitted to hospital suffering from bronchitis. He was discharged on 11 February and returned to his unit on 12 February; on 15 February he was transferred to the Engineers Training Depot at Brightlingsea. That day, Major F W Pugh (27) signed the autograph book but cannot be further identified.

9

Wright in France

His record of service shows that on 3 April Wright proceeded overseas to France via Southampton, and marched in to AGBD at Rouelles on 4 April; however, 2nd Lieutenant **Lester Ferrier** (23) of 1 Pioneer Battalion signed the book on an illegible date in March in Le Havre, France.[12]

Wright marched out to 1 Division Engineers Headquarters on 10 April when he met Captain **John Edward Graham Stevenson**, MC (22), the Adjutant of 1 Division Engineers since 2 February, who signed the book on 10 April. Wright's posting to 1 Division Engineers HQ was in fact a detachment for a limited period from 1 FCE, where he had been taken on strength on 11 April, and was then detached immediately.

The detachment to 1 Division Engineers HQ was no picnic. The unit was initially located at Bertangles, but moved to Eecke and then to Le Grand Hasard, where on 13 April Wright was supervising the digging of support lines in the second zone in front of Morbecque with 200 Labour company men and 400 Chinese; this work continued the next day.

Lieutenant **Cyril Lawrence**, MC (9) signed the book on 16 April 1918. At the time he was serving with 1 FCE, having returned from leave in England on 11 March.

At the beginning of April, Major Thomas Murdoch[13], formerly OC of 1 FCE, was transferred to 1 Pioneer Battalion and Captain Bruce Sheehan Dowling[14], MC temporarily took over the Company; Lieutenant Frank Rochester[15] was in charge of section 1, Lawrence section 2, Lieutenant Robert Osborne Wrightson Earle[16] section 3 and Lieutenant George Richard Thompson[17] section 4. Wright arrived on 11 April but was immediately detached to 1 Division Engineers HQ. On 19 April, Major **David George Sinclair** (42), who signed the book on 24 September, joined the company as Officer Commanding. He had previously been with 3 FCE. On 23 April Lieutenant Rochester was killed in action; on 24 April, Captain Stevenson returned to 1 FCE from 1 Division Engineers HQ, to be replaced there by Captain Dowling; on 26 April, George Knight Kerslake and **Henry Francisco Hetherington** (14) (who did not sign the book until 14 September) were taken on its strength from reinforcements and on 28 April Wright returned from his detachment. So, by the end of April, the officer's corps of 1 FCE included Stevenson, Hetherington, Kerslake, Lawrence and Wright, as well as Sinclair (OC), Earle (no. 3 Section), and Thompson (no. 4 Section). Hetherington, Kerslake, and Wright were all from EOTS Roseville.

Garnet Halloran (left) and Lieutenant Cyril Lawrence (AWM3794854).

During the course of May 1 FCE continued its activities from 1 May – the preparation of a defensive position on front and support lines and of villages as defended localities, of roads for 'blowing [up] in the event of enemy progress along the front', and of cellar accommodation for use as headquarters for battalions and company and for regimental aid posts.[18] One section of the company was allotted to each battalion of 1AI Brigade; at any one time, there were three battalions in the front line, and one in reserve; the reserve section (no. 4 in the case of 1 FCE) was engaged at this moment in the salvage of Royal Engineers stores from the local Strazeele dump.

The company was relieved about every eight days. During the periods of relief, the company undertook a general training program in the morning each day, followed in the afternoon by recreational training and perhaps a bath and new underwear. On 11 May the morning training program included gas training, musketry exercises and section drill, route march and reconnaissance; on other days, there was also work in knotting and lashing, concealment of trenches, and company drill.

On 15 May no. 2 section under Lieutenant Kerslake proceeded to carry out work on camp improvements for 1 AI Brigade, returning on 18 May. The company returned to the front line on 19 May, relieving 3 FCE. The war diary reported that the – *Usual defensive works are being carried out by the sections attached to the line battalions – chief attention being paid to wiring and improvements to existing trenches.*[19]

During the week ending 25 May the company laid 2435 yards of wire, excavated 440 yards of the front line and 1650 yards of the support line, maintained 6 mines, and strengthened accommodation and installed gas-proof doors on several buildings for battalion and company headquarters. A considerable amount of material was salvaged from the Strazeele dump. The company lost 1 other rank killed and one wounded. The man killed was Louis Wilton Poole[20], a colleague of Wright's at EOTS and a member of the July 1917 reinforcements who had embarked aboard the *Euripides* on 31 October 1917 under the command of Edward White.

The company's next period of relief started on 27 May, but this time it took over work from 2 FCE on the support line, consisting of general engineering works for defensive positions. The list of work undertaken during this period reads more like the work it undertook in the front and support lines from 19 May. The company's work was hampered by shortages of material and it was necessary on some occasions to use salvaged material entirely. The work was not all constructive: a large barn that was a ranging mark for enemy artillery – that is, it was used to improve the accuracy of enemy shelling – was demolished by gun-cotton explosive.

On 5 June sections 3 and 4 under Lieutenants Earle and Hetherington moved to Ebblinghem for work and a composite section under Lieutenant Peterson moved to work with 1 Brigade headquarters. The list of work completed in the period ending 8 June includes wiring, trench excavation and draining, building of breastwork, parapet and parados, completion of machine gun positions, and sinking of mine shafts. On 11 June Lieutenant Kerslake went on a course at 2 Army Gas School.

Engineer work included demolition as well as construction, and the June 1918 war diary includes a detailed description of the destruction by explosive of a bridge and a crossroads on the night of 12/13 June. The bridge was completely demolished by several charges, each of 66 pounds of gun cotton. The crossroads charge consisted of 200 pounds of ammonal and, according to the war diary, 'the crater formed was 40 ft in diameter and 19 ft deep.'[21]

This demolition work was in preparation for an enemy attack expected in the next few days. In the meantime, apron fences were erected, trenches excavated and improved, breastwork built, and crops cut for field of fire. On 17 June, Lieutenant Kerslake returned from gas school and Lieutenant Earle went on 14 days furlough in England. On 20 June Lieutenant Hetherington went to hospital, accidentally injured.

On the night of 23/24 June, in a local operation, two parties of 10 sappers each from 1 FCE assisted 2 Battalion in the work of consolidating wiring. They encountered little resistance, but platoons from 3 Battalion sustained heavy casualties from enemy artillery fire, but reserve troops sent in helped gain the objectives. Relieved on 27 June, the company moved back to rest billets, and Lieutenant Kerslake and his section went to Ebblinghem to commence sundry jobs at Reinforcement Camp; Lieutenant Peterson started giving instruction in the Lewis gun to sappers. He continued the next day.

After the war, Wright reported that he had been gassed in his eyes in June 1918 in France and had suffered a burn to a lower lid. The 1 FCE war diary makes no mention of a gas attack on the company at that time. The later medical report included mentions of two other optical conditions that affected his eyesight, but the conclusion of the medical board considering the case was that, though the disability complained of was due to military service, the present degree of disability was not permanent and a complete improvement could be expected in six months.[22]

The July war diary revealed that, at the start of the month, the company was split up, with members in several different locations. Headquarters and sections 1, 3 and 4 were located together, undertaking work for brigade as required; Lieutenant Kerslake was still in Ebblinghem with section 2, not to return until 4 July; and the mounted section was supplying transport as required.

On 5 July the company moved back into the front line. No. 1 section (under Lieutenant Wright) was working with 1 Battalion on the right front; no. 2 (Lieutenant Kerslake) on bridge and road mines and salving from the Strazeele dump; no. 3 (Lieutenant Earle, back from his furlough) on several different tasks; no. 4 section (Lieutenant Peterson) with 4 Battalion on the left front. On 7 July, the OC (Major D G Sinclair), Captain Stevenson and Lieutenant Wright were siting posts for Strazeele station defences and inspecting Moleghein Farm defences on the right. On 11 July Lieutenant **Frank Granville Moffitt** (10) was taken on strength. He signed the autograph book on 25 July 1918. On 13 July a party of engineers from the American Expeditionary Force were attached to 1 Brigade for experience; two officers and 6 NCOs subsequently arrived with 1 FCE for two days' experience. At the end of the month the company moved from Borre to Racquingham. **Allan M Brown** (30) and **J C Greenwood** (30) of the Young Men's Christian Association signed the book on 26 July 1918; the reason for their visit is not set out in the war diary. Lieutenant **Alexander Jarvie Hood Stobo** (41) signed the book on 30 July 1918; he was in the process of moving back from an attachment to 119 Infantry Brigade to his home unit, 1 Battalion, which was at Racquingham for recreation.

Australians walking through Harbonnieres August 1918. Australians walking through the ruins of Villers-Bretonneux August 1918 (AWM E02193)

The company Concert Party gave two performances on 3 August, which the war diary described as 'good' and 'largely attended'.[23] The CRE and officers of 1 Brigade were present at the evening performance. Their reaction is not mentioned.

On 5 August there was a major re-organisation of the company with the re-assignment of many of its Lieutenants. Lieutenant Moffitt went on a billeting party to scope out the next location for the company's billets; Lieutenant Earle left for Australia on furlough; Lieutenant Lawrence and one OR went on a course of training at R E Rouen; and Lieutenant Hetherington took charge of the transport section. On August 6 the company moved to St Ouen, thence to Rivery, ending up at Villers-Bretonneux on 9 August, and, to keep an appropriate distance from the front line, the sappers were billeted in Harbonnieres and the transport in Bayonvillers on 10 August. The enemy started bombing and artillery attacks on both locations and kept them up for the next day as well. Lieutenant Moffitt and 3 ORs were wounded. Following a decision by 1 Brigade to adopt a policy of active defence, the assistance of sappers was required on wiring and improving their front lines. During this operation, an old British front trench was discovered that had a number of saps extending forward that were a menace to the garrison and needed to be shut off. Lieutenant Peterson and no. 4 section attended to those on August 12; Lieutenant Kerslake and no. 2 section worked on improving the front trench occupied by 2 Battalion. Captain **William Edmund Fitzgerald** re-joined the company on 14 August after a long period of absence.[24]

On 15 August the enemy made two bombing raids near the company's billets, and artillery was active, so the company moved on 16 August to Vaux-sur-Somme, the sappers via Bayonvillers and Hamel and the transport via Villers-Bretonneux, for a rest period of 5 to 7 days. At the end of the period, on 23 August, 1 FCE took part in an operation to capture Bapaume. Captain Fitzgerald was liaison officer with 1 Brigade for this operation. Amongst the company's activities were the preparation of a building as an aid post and the reconnaissance of water supply on the left front of the Brigade. This work was carried out by Lieutenant Wright in charge of the section.

No. 2 section under Lieutenant Kerslake constructed dugout for 3 Battalion headquarters and carried out reconnaissance of country captured in the advance and reported on water supply, enemy dumps, railways and rolling stock, and dugouts.

On 25 August Lieutenant Wright left for a 14-day furlough in England, followed on 26 August by Lieutenant Kerslake. On 27 August Lieutenant Higgins re-joined from Brightlingsea in England. Lieutenant Hetherington took charge of no. 1 section.

Major **Henry Hunsted** (21), who signed on 28 August 1918, was in the Royal Army Medical Corps and OC Troops aboard HM Hospital Ship *Newalia* but cannot be further identified. It is possible that Wright travelled from France to England aboard the *Newalia*, and they met then.

By early September, after a period of rest, the company moved to Mt St Quentin where it was engaged in burying the dead and improving sanitation. It moved to Boucly on 9 September and then on to the front line. On 11 September, the day Lieutenant Wright returned from furlough, Lieutenant Peterson was killed in action, and was buried the next day.

Lieutenant **Samuel McConnel** (26) signed on 12 September 1918. He was in charge of no. 1 section of 1 FCE.

Lieutenant Kerslake returned from furlough on 13 September, when the company was employed on improving trenches, accommodation and reconnaissance.

Lieutenant **Walter Henry Higgins** (26) signed on 14 September 1918. That day, the war diary noted that both of the company's locations had been heavily shelled due to their proximity to batteries. The company moved to billets near Brusie, except for no. 2 section, which was constructing advanced brigade headquarters.

Captain Fitzgerald (26) signed on 15 September 1918.

On 15 September the company was resting and improving billets. At about 4.00 pm Lieutenant Kerslake and Captain Davidson (of 2 FCE) were in the officers' mess of 2 FCE near Roisel when the enemy opened a barrage of high explosives. One shell hit the mess, shattering it and killing Lieutenant Kerslake and Captain Davidson, and others. Their death was believed to have been instantaneous. Lieutenant Kerslake's body was buried in Tincourt Military Cemetery the next day.[25]

The dugout in which George Kerslake was killed. AWM P03483.027

The impact of Kerslake's death on Wright can only be imagined. As already noted, they were close, having trained and embarked together, and it is possible that they had met up in England on their recent furlough.

It is also likely that Kerslake had taken the opportunity of his furlough to visit his family in England: his next-of-kin, his father, lived in London. The contents of his record of service include a letter that says that he became engaged to be married in England, but whether he did marry is unknown. There were two family notices of his death published in newspapers in Western Australia describing him as 'sincere friend of Audrey V Palmer of Perth.'[26] Given that Audrey was about 10 years Kerslake's junior it is unlikely to have been a romantic attachment.[27]

Whatever Wright's reaction to Kerslake's death, the war continued, and the day after, Major Sinclair, Captain Fitzgerald and Wright reconnoitred the forward area preparatory to taking over the works there. On 17 September the company took part in an operation that aimed at the outposts of the Hindenburg Line. Its role included the reconnaissance of water supply, repair of wells, searching for enemy mines and booby traps, and accommodation. On 18 September Wright acquired, by means unknown, a Steyr pistol and pouch which remained in the family's collection until recently. The company's work continued for several days, whereupon the company moved to Monfliers for relief and recreation.

The war diary for October notes that the YMCA had provided a reading room and piano to the company, possibly as a result of the visit of Messrs Brown and Greenwood of the YMCA in late July. On 3 October 25 Other Ranks eligible for leave to Australia paraded and were addressed by Captain Stevenson, himself eligible as a Gallipoli veteran. They marched off, amidst much cheering. The company was engaged for the next 10 days in training and recreation, leading up to a parade on 14 October at which the sections were inspected by Major-General T W Glasgow, CB, and Lieutenant Colonel W A Henderson. T/Major C S Steele[28] MC joined the company as OC on 14 October; he signed but did not date the autograph book. The rest of the month was taken up with training and preparation for the sports day scheduled for 29 October, and the establishment of an education scheme. The departure of the 25 ORs for return to Australia earlier in the month, so many of whom were senior NCOs, meant that it was necessary to appoint younger soldiers in their place. To improve their efficiency and give them great confidence an NCOs' class was formed. On 31 October the topics of instruction were Communication Drill and Practical Demonstration in Squad Drill.

November started with preparations for a Review of the Division involving a parade before senior officers and planning for the education program. All these were cancelled or postponed indefinitely when instructions were received to move to the front line. After several false starts and a long journey, the company reached Bazeul on 11 November, to receive news of the Armistice.

Things relaxed after that. There was some thought that the company might be sent as occupation troops to Coblenz in Germany, but this did not eventuate; the educational program was re-established, this time with 2 and 3 FCE involved as well; recreation and sports were pursued earnestly, 1 FCE being very successful in various football matches, although a Divisional Engineers team lost to a 3 Battalion team on 30 December in the presence of the Prince of Wales.

On 9 December Wright returned from leave in Paris; it is not recorded either in the war diary or Wright's record of service when the leave started. On 8 January 1919 Wright went on leave to the UK with the requirement to report to AIF Headquarters at the end of the leave. This was the end of his time with 1 FCE. He duly reported as required, and on 22 January he was attached to the Staff Director of Repatriation and Demobilisation.

On 9 December 1926, Major-General Julius Henry Bruche[29] wrote the following commentary on Wright –

After the Armistice he was employed in the Demobilization Department which was under General Sir John Monash. During this period, while head of one of the sections of the Demobilization Department, I selected him to organise and control a section of the Industrial employment branch of all members of the AIF who were on non- military employment under the Demob. Department. He also carried out the supervision of those members in the Engineering and kindred trades.

From personal experience I know Captain Wright's work was most successful and carried out in an able manner.

He is an officer of high principles and possesses initiative, resource and force of character.

After his return to Australia he was appointed to command the 2nd Field Park Company in the 2nd Division and is now on the Unattached List.

Over the next few months, Wright's book was signed by –

Lieutenant A Carroll (24)
4 ABSRO Company, who signed in London on 12 March 1919 has not been able to be identified
Captain Harold Francis Darby (27)
6 Battalion, signature undated, an accountant of Hawthorn, Victoria
Staff Nurse **Josephine Margaret Gannon**(13)
Australian Army Nursing Service, signed in London on 15 May 1919
Major **Howard George Tolley** (18)
Formerly CRE, 5 Division, signed on 2 July 1919
Lieutenant **Frank Roland Morris** (41)
Formerly of 1 Pioneer Battalion; about to start a period of non-military employment, London on 9 July 1919
Sister Ethel Brice Butler (23)
AANS, undated but timing matches, nurse, of Kilcoy, Queensland
Lieutenant Hector Lee (23)
33 Battalion, undated but timing matches; undertook a period of non-military employment in 1919; indent agent, of Melbourne, Victoria
Lieutenant James Newson McAnna (10)
43 Battalion, undated but timing matches; undertook a period of non-military employment in 1919; engineer, of Adelaide, SA
Captain William Lauchlan Sanderson, OBE CdeG, MC (21)
4 Division Ammunition Column, undated but timing matches; undertook a period of non-military employment in 1919; surveyor and drainage engineer, of Claremont, WA

In due course, Wright took non-military employment leave with pay and subsistence allowance from 20 August to 10 September; the leave was extended to 12 December, and he was permitted to travel to France on 3 November to spend the remainder of the leave in Paris, Nice and Brussels to obtain further experience in manufacturing goods. Subsequently the leave was extended again, to 10 January 1920. On 22 January he embarked aboard the S S *Friedrichsruh* for return to Australia, with the duties of an education officer.

Wright and Durrant returned to Australia aboard the *Friedrichsruh*. This was not their first encounter: Durrant had been appointed Deputy Director, Branch ID Repatriation and Demobilization Department, AIF Headquarters, London, on 11 December 1918. As such he was Bruche's superior, and Bruche, as we have seen, was Wright's superior. Bruche and Durrant had to deal with the matter of Palmer, who resigned his commission and sought to be paid money supposedly due to him whilst he was on non-military employment leave. This would have been contrary to Ruling 27, and the enforcement of Ruling 27 was required by the Director General of Repatriation and Demobilization, General Sir John Monash. So Palmer missed out, but the decision came from the top.

Several of those who signed did so aboard the SS *Friedrichsruh* as they returned from England -

Captain **George Lancelot Allnutt Thirkell** (34)
3 Division AE, 8 February 1920
Lieutenant **Charles Jabez Jennings** (2)
46 Battalion AIF, Great Australian Bight on 5 March 1920
Lieutenant James William Allan (37)
3 Division Signal Company AE, 8 March 1920; journalist, of Ballarat, Victoria
Lieutenant Reginald Edward Clarke (29)
5 Battalion, 8 March 1920; cashier, of Deepdene, Victoria
Captain Cecil John Fletcher (34)
Headquarters, Australian Corps, 8 March 1920; telegrapher, of Haberfield, NSW
Marion L Hunt (17), civilian   8 March 1920
John Lawson (63), unidentified 8 March 1920
Captain **Lyndsay Torrance Maplestone** (17)   8 March 1920
Captain Daniel Alexander Clyde McNicol (10)
AAVC, 8 March 1920; veterinary surgeon, of St Kilda, Victoria
Captain Harris Mendelsohn (9)
AAMC, 8 March 1920; medical practitioner, of St Kilda, Victoria

T Pateman (29) Cannot be identified, 1920
A N W Root (9) Cannot be identified, 8 March 1920
Edgar Ainslie (42), civilian, YMCA representative, undated
Lieutenant Allan Israel Brown (20)
27 Battalion, undated; Methodist minister, of Eastwood, SA
Lieutenant Colonel James Murdoch Archer Durrant, CMG DSO (36)
Deputy Director Repatriation and Demobilisation Branch, AIF HQ, undated; soldier, of Victoria Barracks, Brisbane, Queensland
Captain Alexander Fraser, MC (16)
38 Battalion, undated; soldier, of Essendon, Victoria
Lieutenant George Edward Gaskell, MC (38)
1 Battalion, undated; clerk, of Fitzroy, Victoria
Captain Llewellyn Griffiths (17)
17 Battalion, undated; formerly Commander, Royal Navy, of Sydney, NSW
Henry Francis Trafford Heath, Chaplain (28)
Signed on the Suez Canal, undated; minister of religion, Adelaide, SA
William Sherwood Houghton (21)
CCMO, 32 Battalion, undated; barrister and solicitor, of Bairnsdale, Victoria
Captain Arthur Poole Lawrence, MC (9)
AAMC, undated; medical practitioner, of Ascot Vale, Victoria
Captain John Loughnan, MC (42)
59 Battalion, undated; civil servant, of East Melbourne, Victoria
Lieutenant Colonel Donald Ticehurst Moore, DSO CdeG CMG (21)
3 Battalion, undated; clerk, of Randwick, NSW
Lieutenant Arthur Leslie Murray (34)
15 D U S, undated; clerk, of Brisbane, Queensland
Lieutenant Thomas Orr (42)
12 Battalion, undated; farmer and orchardist, of Launceston, Tasmania
Captain Gunning Francis Plunkett (38)
3 Battalion, undated; auctioneer, of Yerong Creek, NSW
Lieutenant Gavin Michael Quigley, MM (21)
7 Field Artillery Brigade, undated; grazier, of Warren, NSW

Captain James Edmund Savage, MBE MC (19)
6 Australian Division Train; soldier, of Melbourne, Victoria
Colonel William Floyd Shannon, OBE Chaplain (20)
Senior Chaplain (Presbyterian), undated; clergyman, of Maylands, WA
Lieutenant Erith Valentine Stratton (36)
1 FCE; promotion to Lieutenant was on 2 January 1919; draftsman, of Hurstville, NSW
Colonel **W H Tunbridge**, CB CMG VD OBE (36)
Director of Mechanical Transport Services
Major Clement Robert Walsh, MC (9)
3 Divisional Train; pharmaceutical chemist, of Newcastle, NSW

Some of those signing are known, but either the place or time of signing or both is not -

Lieutenant Charles Alfred Hopton (17)
8 Field Company, undated; engineer, of Adelaide, SA. Hopton was a contemporary of Wright's at EOTS. He was commissioned 2nd Lieutenant two months before Wright and embarked for France several months before. He was the last officer of 8 FCE before it was disbanded in 1919
Frank J Hubbard (21)   Education service
Lieutenant Charles Reginald Palmer (37)
5 Pioneer Battalion
Lieutenant Colonel Ian T Watson (38)
New Zealand Expeditionary Force

Some are not known and cannot be traced –

Captain James Blair Donaldson (34)
RAMC, undated
Lieutenant D H A Gill (12)
Engineers, undated

Lieutenant M A Mollemodge (34)
Royal Navy
Lieutenant J Ponton (21)
32 Battalion

The voyage was not without incident. Amongst the passengers was a group of former Australian soldiers who had been discharged in England and offered third-class passage aboard the *Friedrichsruh* back to Australia. They found the conditions intolerable, and protested strongly to the military authorities on board, who informed them, not unreasonably, that they had no jurisdiction as they were civilians. They found that also on their decks were prisoners from the AIF detention barracks at Lewes, and VD patients, both of which groups had the run of the ship. They were required to draw their own rations, appoint their own mess orderlies, wash up their own utensils, and clean the benches on the mess deck.

During the vessel's visit to Colombo – its first stop since leaving England - there were allegations of serious misbehaviour made against some of the men on board towards hotel staff and other civilians.

During the visit to Fremantle some of the prisoners escaped and were not re-captured. In addition, there was a fracas aboard at the dock in Fremantle, and the crew walked off the ship, only returning after the Commonwealth agreed to indemnify the crew against any damage that occurred during the rest of the voyage.[30] The *Argus* reported that the *Friedrichsruh* would be 'the last troopship carrying returning soldiers who will be brought through the city [Melbourne] to the Sturt Street depot.'[31] The ship arrived in Sydney on 12 March.

On his arrival in Sydney in March Wright underwent a medical examination for an eye problem, a growth over his right eye. As noted above, the medical report attributed it to his military service but found that the degree of disability was not permanent and expected that there would be complete recovery in six months; he was fit for work, and for discharge from the AIF. This occurred on 12 September 1920.

10

After the War

Alfred certainly hit the ground running when he disembarked from the *Friedrichsruh*. The first matter to deal with was business, and then there was a family one. Family first.

Family Matters

It cannot have been easy for couples to re-establish their relationship upon returning from the long absence of the war. Alfred had left at the end of October 1917 and returned after 2 ½ years. It is possible that Charlotte resented his absence, leaving her with a young family, and the extension of his time away caused by his non-military employment leave in Paris, Nice and Brussels. Alfred took legal action, seeking a decree for the restitution of conjugal rights on Charlotte's part. *The Northern Star* reported that he said she –

had objected to his going to the war because her mother was a German. While he was in England, he received a cable from her saying she did not intend to live with him again. She accused him of kicking and beating her, and of infidelity. He denied he had ever been cruel to her or had ever given her any other cause to leave him. "There is another man in the case," said [Alfred]… An order was made directing [Charlotte] to return to her husband within 21 days of service upon her of the decree.[32]

Charlotte did not comply with the decree and a decree nisi was granted in March 1921. The decree was made absolute on 31 October 1921. Charlotte married Edward Burge within a few weeks.

Business Matters

At the end of March 1920 Alfred established A R Wright and Company, merchants. The company's offices were on the first floor of Bridge Street Chambers, 17 Bridge Street Sydney. In the same building were other firms of merchants, hardware brokers, importers, manufacturers' agents, electrical engineers, even an accountant. Not long after, on 22 May, A R Wright & Co imported one case of wire from England, and on 15 June, two cases of lamp-ware, also from England, so no grass grew under Alfred's feet when it came to business. Nor to creative work, either: on 2 June there was notification of his application for a patent for a 'Trousers creaser and hanger and skirt-hanger' lodged between 6 and 12 May.

Reports of imports by AR Wright & Co later in 1920 included 6 cases of trimmings (a drapery product) and 12 cases of paste gum; in 1921, more trimmings, and toys; in 1923, 2 bales of canvas.

Alfred was also selling second-hand cars and motorcycles, as well as verandah blinds, tarpaulins and deck chair canvas, lamp shades and children's winter slippers. The company advertised throughout the north coast and northern tablelands offering to buy second-hand woodworking, engineering and mining machinery either outright or on commission, and inviting buyers or sellers to send in particulars of machinery they wanted to dispose of or purchase. A R Wright and Co were also sole Australian representatives of Grand Marnier Liqueurs.

The company's first advertisement in the 1921 Sands Directory was a simple 'Wright AR and Co 17 Bridge St' in the 'Merchants – General' classification. It was the same in the 1922 edition. In 1923, however, things had moved on, better and brighter – there was a telephone number and a cable address ('Wrigo, Sydney'), and the descriptor 'General Merchants and Manufacturers.' There was also an advertisement under 'Manufacturers and Importers.' In 1924, the advertisement in the 'Merchants – General' classification was better still - a new address ('Ordnance Buildings, Circular Quay, Sydney' – more specifically, 148 George Street), a showroom at 164 Pitt Street, and a much more detailed description of what the company offered:

General Merchants and Manufacturers of Axe, Pick and Implement

Handles, Canvas Goods, Verandah Blinds, Holland Blinds, Horse Rugs, Tents, Tarpaulin, etc, Knitted Goods (silk and wool), Hosiery, Costumes, and Artistic Lamp Shades, etc. Show Rooms, City House, 164 Pitt St.

It was to the establishment here at 164 Pitt Street that newspaper advertisements directed potential customers from the beginning of 1924. There was a somewhat smaller advertisement in the 'Manufacturers and Importers' classification.

In the 1925 edition, there were two telephone numbers, a telegraphic address as well as a cable address, and showrooms at City House and at 250 George Street, near Bridge Street, in both the 'Merchants – General' and the 'Manufacturers and Importers' classifications. This was a company on the way up.

Outside the office, Alfred maintained his interest in military matters. He donated a prize for shooting at the first annual prize meeting of the 1 and 2 Division Field Engineers in 1923.

In 1924 Alfred applied for the extension of the terms of a special lease over 58 acres of land in the parish of Roseberry, near Casino, for a sawmill site and tramway. The term was extended from 1 January 1925 to 31 December 1927 for an annual rent of £26. The amount payable by 31 December 1924 was £25/16/8, taking into account the deposit already paid and expenses charged. On 11 February 1925 the Casino Pastures Protection Board told Alfred that it had no objection to a special lease of this land, which was held in the interests of travelling stock.

But it was all about to fall apart. A bill of sale holder appointed PB Allpress as receiver, who invited all persons having claims against the estate of AR Wright and Co to forward particulars to him at the Ordnance Building by 4 March. On Saturday 28 March the Bailiff in Kyogle gave notice that goods and chattels of A R Wright and Company, comprising timber, logs, sawmill and plant and machinery would be sold by auction to pay debts to Digby and Eva Lavelle. Things got a lot worse: on 9 May FA Lindsay, Public Accountant and Auditor, offered for sale 'assets in the assigned estate of Alfred Richmond Wright, Merchant,' tenders to close on 14 May. AR Wright and Company was bankrupt.

The assets of Alfred's estate included stock and plant of the knitting factory (valued at £1546/15/3), hardware and ironmongery stock, including some motor accessories (£1272/18/6) and office and household furniture, fittings, partitions, etc (£252/6/-), and the lease of premises. The location of the goods was shown as 'Ordnance Building, Hickson Road, off George Street North, near Circular Quay.'

In the event, the whole of the stock, plant, office equipment, etc, was put up for auction on Friday 29 May. There was so much to be dealt with that the sale continued into the next week. The final distribution from Alfred's bankrupt estate was not made until November 1926.

On 13 April 1927 Alfred retired from the unattached list of AIF officers with the rank of Captain, and permission to retain their rank and wear the prescribed uniform.

Alfred returns to the business fray

On 1 April 1928 the *Iron Chief* went aground on Mermaid Reef, near Crowdy Head on the NSW north coast, with its cargo of 11,227 railway sleepers valued at £4000. After frantic efforts by a local tug, she came to rest on Diamond Head beach. On 17 April the hull and the cargo were sold, the ship for £160 to Mr E S Budrodeen, a shipbreaker of Drummoyne, and the cargo for £390 to a syndicate that included Alfred, Budrodeen and two others. The syndicate had a month to remove the sleepers from the wreck, which was lying on sand with its starboard side awash in rough weather and 3 feet of water over the cargo. The challenge to recover the cargo included the difficulty of removing it, and the weather.

In August the Public Works Department of the Commonwealth Government called for tenders for the repair of the roof of the Ordnance Stores. The initial round of tenders was referred to the Works Director in Central Square in Sydney, reason unknown, but possibly for either detailed consideration of the prices received, or consideration of Alfred's capacity to undertake the works, or determination that Alfred did not have an advantage over other tenderers by virtue of his occupation of the building.[33] In September it was announced that Alfred had won the tender.

Wreck of the *Iron Chief* on Mermaid Beach.

Alfred opened the business he called the Quay Bargain Store at 148 George Street on 19 September and registered it on 24 September 1930.

In 1931 Alfred won the tender to build brick additions to a building at Burlington Road, Homebush.

11

Alfred's Death

Alfred died suddenly in Sydney on Thursday 17 March 1932 at the age of 42. The Northern Star, in Lismore, reported that 'the late Mr. Wright was spending the evening with some friends when he collapsed and died in a few minutes.'[34] He was interred in the Church of England cemetery at Rookwood on 19 March, with the family and friends of his children and mother and sisters, as well as the officers and brethren of Prince Alfred Lodge no. 94, invited to attend the funeral.[35]

On 24 March, Daphne Wright, Alfred's elder daughter, wrote to 'Defence Department, Melbourne' following up on an earlier telegram asking whether it had Alfred's will. She wanted it to 'institute a thorough search into [its] records' to try to find the will. According to the letter, which is preserved in Alfred's record of service, Alfred had 'left certain business, the transaction of which I am unable to carry out owing to the absence as far as so far can be seen of any will... on his departure for France during the currency of the Great War, a Will was left which should now be in your office. The finding of this Will would greatly facilitate the continuance of his business, which I have taken over.'

The search that Daphne asked for revealed that the will had been forwarded to 2 Military District [NSW] on 23 February 1920 for return to officer [i.e., Alfred]. An undated statement on file by the officer said 'copy lodged with wife,' which is confirmed by a certificate signed by Alfred and attached to his record of service. The officer responding to Daphne's request reported this in their response to her letter, adding a suggestion that, as a copy of the will might have been in the possession of the District Finance Officer at Victoria Barracks, Paddington, she should communicate direct with that officer.

The next relevant document is not on Alfred's record of service, but on that of James Alexander Hay. The NSW Public Trustee wrote requesting the current addresses of the witnesses to the will, James Alexander Hay and George Percy Ross. This was in response to a letter from the Public Trustee saying that it was necessary for it to obtain an affidavit from one of the witness as to the execution of the will.

It turns out that Alfred executed a will on 3 April 1918. That day, Alfred, James Hay and George Ross had left Brightlingsea to proceed to France via Southampton, and there must have been a feeling amongst them that it would be a good idea to have made a will. Alfred verified George's signature on the certificate as to the location of George's will.

When administration of Alfred's estate was granted to the Public Trustee on 6 December 1932 the will was annexed.

In May and June FR Strange, auctioneers, held sales on behalf of the Public Trustee to clear the estate of goods, but Christie's auctioneers still had goods to sell in October 1933.

12

Extended Bibliographical Notes

Francis William Broughton

Lieutenant Francis William Broughton (36) was a dentist from Woollahra, NSW, who had enlisted on 22 October 1917 and, when he disembarked in Devonport on 26 December, was assigned to Parkhouse no. 2 Camp and then to Sutton Veny.

He did not arrive in France until 2 December 1918.

He embarked aboard HT *Anchises* on 22 August 1919 for return to Australia. His appointment was terminated on 12 November 1919.[36]

Allan M Brown and J C Greenwood

Allan M Brown (30) was the Secretary of the YMCA and J C Greenwood (30), the Military Secretary. Both had been active early in the War in raising funds for troops both at home in Australia and in France, and they had travelled to France to assist in YMCA operations there. This work was regarded very positively at home and overseas –

There is no branch of work which is being undertaken in the interests of soldiers at the front that stands out so nobly as that of the Young Men's Christian Association. From the commencement of the war this admirable institution has, first in camp at home, then on the troopship, again at ports of call, and in England and France, supplied those slight home and social elements which have done so much to relieve the loneliness and homesickness and fill the idle hour of the men of our battalions. No sect or creed is known. All a man has to do is to behave himself, and he is welcome to all the comforts and little accessories which the YMCA provide. At all our Australian camps the YMCA tent has always been most popular. Social evenings, manly sports, and occasionally an effort to reach the wayward have characterised the work of the agents of the association. On the troopships each soldier has been supplied with writing paper and envelopes and with magazines and literature for his use. In port at Cape Town or Durban there is always a cheery welcome during the few hours ashore, and so again in Egypt, London, and at Salisbury Plains the soldier meets the cheerful YMCA tent with its hearty invitation and welcome to make it his club for recreation and rest. But the YMCA goes further than this. It strives to protect the men from those evils which dog the footsteps of soldiers, and many a mother can rest more contented in Australia to-day when she knows that her boy is being safely guarded and advised by men who know the evils that lurk ready to ensnare him.*[37]*

Gerald Mosman Carr

Captain Gerald Mosman Carr (27) was an architect in Sydney when he enlisted on 5 May 1915 in 24 Battalion.

Carr served with distinction at Gallipoli and Pozieres, being mentioned in dispatches for good services rendered on 21 December 1916. By mid-1917 he was undergoing medical treatment for a variety of symptoms and in the latter part of the year he was transferred to Parkhouse no. 3 Camp to await assignment to a unit, which is where he met Alfred. His medical condition did not improve, however, and on 2 May 1918 he was transferred to the Permanent Supernumerary List and posted for duty with AIF Depots in the UK.

He married Winifred Mary Wells-Cole on 30 January 1919 and on 3 June 1919 was appointed MBE. He returned to Australia on the *Friedrichsruh* on 22 January 1920, which would have given him the opportunity to renew his acquaintance with Alfred. His appointment was terminated on 11 May.

David Christie

Captain David Christie (25), a doctor with the Australian Army Medical Corps from Narrandera in NSW, was assigned to AAMC Training Depot at Parkhouse no. 2 Camp on disembarkation from the *Euripides* and thence to 1 Australian Army Hospital, Harefield.

He proceeded overseas to France on 27 March 1918 and was attached to 2 Australian General Hospital at Wimereux; during the next few months he served with 8 and 56 Field Ambulance, 3 Australian General Hospital at Abbeville, the Australian Corps Gas School, and 32 Battalion as RMO.

He returned to Australia aboard the *Bremen* on 4 June 1919 and his appointment was terminated on 20 August 1919.

Norman Cook

Lieutenant Norman Cook (20), a civil engineer, enlisted on 14 July 1916 and attended the EOTS at Roseville between 31 August 1916 and 1 February 1917. At that time, he was transferred to the Engineers Depot Sydney and given the rank of acting Sergeant from 2 February. He was appointed 2nd Lieutenant on 21 May and embarked aboard the *Euripides* on 31 October 1917 with the July 1917 Reinforcements.

On arrival in Devonport on 26 December he marched into Parkhouse no. 3 Camp, and then to the Engineers Training Depot at Brightlingsea on 18 February 1918. He proceeded overseas via Southampton on 3 April and arrived on 4 April at the AGBD at Rouelles.

On 10 April he marched out to 3 Division Engineers and then, on 17 April, he was taken on the strength of 9 FCE. He was detached between 10 and 22 May for duty with 567 Army Troops Company, Royal Engineers. He was promoted to Lieutenant on 23 May. He remained with 9 FCE until the end of the war, apart from a period of leave in England starting on 16 October.

His appointment was terminated on 25 July 1919 in England. His record of service shows that he had an appointment as Assistant Engineer with the United Railways of the Havana and Regla Warehouses Ltd in Cuba and was due to sail from London on or about 10 July.

Lester Ferrier

2nd Lieutenant Lester Ferrier (23) had been a Licensed Surveyor before he enlisted on 5 June 1916. Ferrier attended Fort Street Model School in Sydney. Assigned initially to the Armidale Depot Battalion, he attended the Engineer Officer Training School between 22 August 1916 and 1 February 1917, and so would have met many of Alfred's colleagues who were at the School during this period. Promoted to 2nd Lieutenant on 1 April, he had embarked aboard HMAT A38 *Ulysses* with 14 Reinforcements 1 Pioneer Battalion at Sydney on 19 December 1917.

After reaching England on 13 February 1918, he proceeded overseas to France via Folkestone on 28 March, and marched into the Australian Infantry Base Depot (AIBD) on 31 March.

Ferrier was taken on the strength of 1 Pioneer Battalion on 28 April 1918 and promoted to Lieutenant on 7 July. The same day he was detached for duty with 1st Division Headquarters; shortly after returning from that attachment on 17 July he was sent to Administrative Headquarters on 19 July for duty. That completed, he returned to France on 25 February 1919, but after a period at AIBD he returned to England to take up a posting as Instructor at the Survey Course in Southampton and made temporary Captain whilst there.

On 4 July he embarked for return to Australia aboard the SS *Norman*. His appointment was terminated on 24 December 1919.

William Edmund Fitzgerald (third from right) with a group of officers.

William Edmund Fitzgerald

Captain William Edmund Fitzgerald (26), who signed on 15 September 1918, was a New Zealander by birth and education. An Inspecting Engineer in railway construction in Victoria prior to enlisting, he had spent 2 years as a Lieutenant in the NZ Engineers Corps and attended an Officers' Training School at Broadmeadows, Victoria. After his appointment as a 2nd Lieutenant on 8 October 1915 he embarked in the 11th reinforcements 2 FCE aboard HMAT A72 *Beltana* on 9 November, bound for Egypt, where he was taken on the strength of 2 FCE on 28 December.

But not for long: on 22 January 1916, his horse rolled on him in a riding accident and he fractured his pelvis. After treatment in Egypt and Britain, and promotion to Lieutenant on 20 February 1916, he was taken on the strength of 1 FCE on 12 October in the field. He took furlough in England in February 1917 and on 20 May he was detached for duty at Engineers Command 1 Anzac, ending with him re-joining his unit on 3 July. On 21 August he was seconded for duty at the Engineers Training Depot, Brightlingsea, and made Adjutant of the

Depot on 24 August. On 7 November he was promoted to Captain, and he remained at the Depot until 10 February 1918 when he was posted to the School of Instruction at Aldershot, where he stayed until 3 March, when he returned to Brightlingsea. On 26 April he relinquished his appointment at the Depot as Adjutant, but on 30 April he fractured his humerus and did not return to duty until 5 July. He rejoined 1 FCE on 15 August.

He returned to Australia aboard the *Orita* on 23 June 1919. His appointment was terminated on 25 October 1919.

William Montgomerie Fleming

Lieutenant William Montgomerie Fleming (3, *below*), a grazier, enlisted on 6 October 1916 and was assigned to the Australian Army Service Corps. In signing the autograph book, he claimed attachment to 28, 29, 30 and 31 Battalions AIF, but whilst he had qualified for a commission there were none available at the time of embarkation. He was appointed a Lieutenant for the voyage only, and he reverted to Driver upon disembarkation. At the time he embarked he was Nationalist MHR for Robertson, having previously served for 9 years in the NSW Legislative Assembly.

Josephine Margaret Gannon

Staff Nurse Josephine Margaret Gannon (13) was a nurse who enlisted on 10 May 1917 and embarked aboard RMS *Mooltan* on 9 June 1917 for Egypt, where she arrived on 19 July. She embarked aboard the *Osmaniah* for Salonika on 12 August, arriving there on 14 August and taken on the strength of 50 General Hospital on 16 August. In June 1918 she became sick, diagnosed variously as dysentery and debility, but after she recovered, she went down with influenza, and it was not until 25 September that she re-joined 50 General Hospital.

On 15 February 1919 she embarked aboard the *Czaritza* for London on leave, where she undertook studies in Domestic Economy at Battersea Polytechnic. It was during this period that she met Alfred. She was promoted to the rank of Sister on 12 July, the same day that she completed her course. She embarked aboard HT *Canberra* for return to Australia on 23 July and was discharged on 15 October.

Henry Francisco Hetherington

2nd Lieutenant Henry Francisco Hetherington (14) was a contractor and builder when he enlisted on 23 June 1916 and was assigned to FCE reinforcements. On 1 July he started his course at the Engineer Officer Training School, Roseville NSW, as an acting Corporal, promoted on 12 September to acting Sergeant. When he completed his course on 9 December, he transferred to the Engineers Depot at Moore Park in Sydney and in due course embarked aboard HMAT A28 *Militades* on 2 August 1917, freshly minted that day as a 2nd Lieutenant.

After disembarking at Glasgow and a period of time at Parkhouse no. 3 Camp and Brightlingsea he proceeded overseas to France on 3 April 1918, being taken on the strength of 1 FCE on 26 April. He accidentally injured his ear on 18 June and did not re-join his unit until 1 August. He was promoted to Lieutenant on 6 August.

He suffered a bout of influenza in March 1919 just prior to embarking for return to Australia aboard the *Ormonde* on 16 June. His appointment was terminated on 31 August 1919.

Walter Henry Higgins

Lieutenant Walter Henry Higgins, MC (26) was a surveyor when he enlisted on 9 December 1914 and was assigned to 4 reinforcements, 1 FCE as a sapper. He embarked aboard HMAT A8 *Argyllshire* on 10 April 1915, and on 2 June 1915 he was taken on the strength of 1 FCE on Gallipoli.

On 14 September he was promoted to Lance Corporal, but he was not at his post: on the 9th he had reported sick, and was diagnosed with malaria, taken to hospital on Malta and then to London. He rejoined his unit, by then in Egypt, on 15 January 1916, and was promoted to 2nd Corporal on 1 March. He travelled with his unit to France where, on 20 July, he was detached for training at Officers' School, England. On 20 October, according to his record, 'having been passed as qualified for commission in Field Companies, [he was] appointed 2nd Lieutenant in the AIF'; he also 'qualified in competitive examination for 1st place for commission.' He was detached for duty at Administrative Headquarters, London on 22 October.

He proceeded overseas to France on 22 January 1917 and rejoined 1 FCE on 26 January; he was detached for a fortnight with CRE 1 Australian Division for duty on Royal Engineers Dumps. On 3 May he was promoted to Lieutenant. On 18 July 1917 he completed a statutory declaration to the effect that he did not wish to make a will, presumably because it might have brought bad luck. Cyril Lawrence (q.v.) was one of the witnesses to his signature.

On 20 and 21 September he was engaged in an action east of Ypres, consolidating a strong point in Polygon Wood under very heavy hostile shelling; he completed the work very satisfactorily and later was awarded the Military Cross for it.[38] On 19 December he was wounded in action, suffering a gunshot wound to his finger, and he was admitted to hospital initially in France and then to 3 London General Hospital. He did not return to his unit until 27 August 1918.

After the Armistice he took leave in the UK and then embarked on 21 May 1919 for return to Australia aboard the *Osterley*. His appointment was terminated on 17 February 1920.

Charles Jabez Jennings

Lieutenant Charles Jabez Jennings (2) was a bank clerk from Victoria who enlisted on 24 July 1915 and embarked aboard HMAT A7 Medic on 20 May 1916 with 3 Reinforcements 46 Battalion. Taken on the strength of 46 Battalion on 28 September, he was appointed 2nd Lieutenant on 27 April 1917 and Lieutenant on 11 September. Wounded in action three times, he survived the war and embarked on 22 January 1920 for return to Australia aboard HT *Friedrichsruh*. His appointment was terminated on 4 May 1920.

George Knight Kerslake

Lieutenant George Knight Kerslake (20), a civil engineer's assistant enlisted on 12 May 1916 and attended the EOTS at Roseville between 20 August 1916 and 1 February 1917. At that time, he was transferred to the Engineers Depot Sydney and given the rank of acting Sergeant from 2 February. He was appointed 2nd Lieutenant on 21 May and embarked aboard the *Euripides* on 31 October 1917.

On arrival in Devonport on 26 December he marched into Parkhouse no. 3 Camp, and then to the Engineers Training Depot at Brightlingsea on 17 February 1918. He proceeded overseas via Southampton on 3 April 1918 and arrived on 4 April at the AGBD at Rouelles.

On 10 April he marched out to 1 Division Engineers and then, on 26 April, he was taken on the strength of 1 FCE. He was promoted to Lieutenant on 23 May. He went on leave to the UK on 26 August and returned on 10 September. He was killed in action on 15 September. (see pages 36, 44 of ROS)

George Hilfers Koch

Lieutenant George Hilfers Koch (27) was a clerk when he enlisted on 25 August 1914 and was assigned as a Private to G Company, 9 Battalion AIF. Things moved quickly as his previous military experience in 1st Queenslanders (Moreton) Regiment was recognised: he was promoted to Lance Corporal on 5 September, skipped Corporal, and to Sergeant the same day, and to Colour Sergeant the next day, 6 September, 11 days after enlisting. It was with this last rank that he embarked with 9 Battalion aboard HMAT A5 *Omrah* in Brisbane on 24 September 1914, disembarking at Alexandria on 6 and 7 December. On 1 January 1915 he was appointed CQMS.

By 2 March 1915 he had embarked aboard the *Ionian* to take part in the Gallipoli campaign, and after work on Lemnos the Battalion landed with the first wave of troops at Gallipoli.

The horrible toll of the intense fighting meant that new officers were required sooner than they could arrive with reinforcements, and on 10 July Koch was one of four former NCOs promoted to 2nd Lieutenant. On 11 August he was sent away ill, eventually ending up in 2 London General Hospital for treatment for a complaint variously described as dysentery, pyrexia, influenza, diarrhoea and gastritis.

He did not return to Gallipoli; indeed it was not until 11 May 1917 that he re-joined 9 Battalion, which was then in training at Bapaume. On 16 May he was promoted to Lieutenant. During the next few months he spent time with the Battalion, in hospital, or on leave, until 17 November when he marched into Parkhouse no. 3 Camp, where he met Alfred. He re-joined his unit on 21 January 1918 and was wounded on 15 March; after recuperation in England he re-joined the unit on 24 September.

He marched out and embarked for return to Australia on 13 October. His appointment was terminated on 23 February 1919.

Cyril Lawrence

Lieutenant Cyril Lawrence (9) was a draughtsman when he enlisted on 11 December 1914 and was initially assigned as a Sapper to 3 Reinforcements FCE. He transferred to 5 Reinforcements 2 FCE on 1 February 1915 and was promoted to Lance-Corporal on 16 April.

He embarked aboard HMAT A20 *Hororata* on 17 April 1915 and landed in Gallipoli to be taken on the strength of 2 FCE on 2 June 1915. On 10 September he was made Temporary Headquarters Sergeant, this appointment confirmed on 4 December.

He moved with 2 FCE to France at the end of March 1916 and was detached for Officers Training School on 20 July; he joined the School of Instruction at the Engineers Training School at Deganwy in Wales on 31 July. *Having been passed as qualified for commission in Field Company*, according to his record of service, Cyril was appointed 2nd Lieutenant on 20 October.

He proceeded overseas to France and was taken on the strength of 1 FCE as supernumerary on 19 December. Here he worked alongside Captain John Stevenson (q.v.). On 12 April 1917 he was attached to Commander Royal Engineers 1 Australian Division for duty, returning to his unit on 11 May. On 3 May he was promoted to Lieutenant, at the same time as Walter Higgins (q.v.).

He spent the rest of the year, brief periods of leave and a 2-week detachment to 1 Division Engineers Headquarters in December excepted, with 1 FCE. He was awarded the Military Cross on 1 January 1918 for his untiring energy, resourcefulness and courage in consolidating a strong point and constructing a communication trench south of Black Watch corner along the Menin Road east of Ypres on 20 and 21 September 1917. He was seconded to the Engineer Training Depot at Brightlingsea on 14 September 1918.

He embarked aboard the *Delta* for return to Australia on 25 January 1919, and his appointment was terminated on 9 May 1919. He re-enlisted for service in World War II and gained the rank of Brigadier.

Lyndsay Torrance Maplestone

Captain Lyndsay Torrance Maplestone MC (17), Adjutant, Australian Auxiliary Transport Company, enlisted on 18 November 1915. After a period in the ranks and training at Duntroon between 26 April and 20 June he was commissioned as 2nd Lieutenant in the 3rd Australian Auxiliary Mechanical Transport Company on 15 August 1916. Promoted to Lieutenant on 7 October 1916, he embarked aboard HMAT A34 *Persic* on 22 December 1916 for Devonport. On the voyage he was appointed ship's quartermaster.

On 20 June 1917 he proceeded overseas to France where, on 25 June, he was attached to the 5th Australian Ammunition Sub-Park near Rouen. He was wounded on 28 September whilst in charge of 15 lorries carrying ammunition at Hell Fire Corner[39] but, after his wounds were dressed, he remained on duty for several hours directing operations and getting all lorries away quickly and safely, even though under very heavy shelling. After being relieved he was admitted to hospital at Rouen. For this action he was awarded the Military Cross.[40] After treatment there and in England he resumed duty on 25 December with 3rd Australian Ammunition Sub-Park. On 12 March 1918 he was posted to 1 Australian Siege Brigade Ammunition Column and from there to the Australian Corps Mechanical Transport Column as Adjutant. On 17 October as temporary Captain he became Adjutant of the Senior Mechanical Transport Director; he finished the war in this role.

After a period of leave, further work in mechanical transport units, promotion to Captain on 18 March 1919 and marriage to Ethel Broad on 22 December he returned to Australia aboard the *Friedrichsruh* on 22 January 1920. His service was terminated on 24 August 1920.[41]

Colonel Walter Howard Tunbridge (front row, centre) and Captain Lyndsay Torrance Maplestone (front row, right)

Alexander Duncan McArthur

Lieutenant Alexander Duncan McArthur (4) was a nurseryman when he enlisted on 31 January 1916 and was assigned to A Company of 39 Battalion. After his arrival in England aboard the HMAT A11 *Ascanius* on 18 July 1916 he proceeded overseas to France on 23 November to join his unit.

He was promoted to 2nd Lieutenant on 17 January 1917. He was wounded in the right hand on 31 January but remained on duty; he was wounded a second time on 8 June when he was gassed but returned to duty the same day. He was promoted to Lieutenant on 23 June.

On 14 August he was transferred to 10 Training Battalion in England, where he attended courses in rifle, Lewis gun and revolver at the Southern Command School of Musketry; he returned to 10 Training Battalion on 30 September. He was admitted to Fovant Military Hospital for an undetermined illness on 2 December 1917. It is not clear when he arrived at Parkhouse, but he left there on 13 March 1918.

He re-joined his unit in the field on 4 June. On 10 September he was wounded in action again, but this time more seriously: a gunshot wound to his left arm that required amputation. After treatment he embarked for return to Australia on 12 December 1918 aboard *Nestor*. His appointment was terminated on 28 May.

Samuel McConnel

Lieutenant Samuel McConnel (26), who signed on 12 September 1918, was a civil engineering graduate from Britain when, on 5 October 1916, he enlisted with the rank of 2nd Lieutenant and was assigned to 2nd reinforcements, 1 Mining Corps, as a member of which he embarked aboard HMAT A16 *Star of Victoria* on 31 March 1916 in Sydney.

He was taken on the strength of 3 Tunnelling Company on 9 July and then, after brief periods in 2 Field Company and 1 Division Engineers in early September, transferred to 1 FCE on 8 September. He was promoted to Lieutenant on 15 January 1917. He stayed with 1 FCE, one bout of sickness excepted, until 4 October 1917 when he suffered a severe gunshot wound to his thigh and was transferred to 3 London General Hospital for treatment.

This occurred towards the end of an operation when he was in charge of the construction of a strong point in the vicinity of the front line. The party came under heavy shell fire but it managed to mark out the post and commence work. He was wounded rather badly but before he left the job made sure that his NCOs understood what was required of them, and then insisted on walking to company headquarters to report on progress made. For this he was recommended for the award of the Military Cross, but the recommendation was not approved.

His recovery meant that he did not return to his unit until 18 January 1918; just after, on 20 January, he went to Corps Gas School for a week, and soon after returning from that he was attached for duty with 1 Australian Infantry Brigade headquarters. On 13 March he was gassed and transferred to hospital in England; he did not re-join his unit until 15 August.

After the Armistice he took 3 months' non-military leave to obtain experience in reinforced concrete work with a company in London. He returned to Australia aboard *Zealandia* on 12 May 1919. His appointment was terminated on 16 August 1919.

Frank Granville Moffitt

Lieutenant Frank Granville Moffitt (10), born in Florence, Italy, was an electrician when he enlisted on 22 August 1914 and was assigned as a private to E Company, 10 Battalion AIF.

He embarked aboard HMAT A11 *Ascanius* in Adelaide on 20 October 1914: 10 Battalion was headed for Egypt and eventually Gallipoli. However, he was transferred as a sapper to 3 FCE on 28 February 1915 and was with the unit when it landed at Gallipoli on 25 April 1915. On 8 September he was diagnosed with influenza and enteric fever, with a recommendation that he return to Australia. He embarked on HMAT A63 *Karoola* on 4 November. He returned to duty on 11 February 1916 with the 4th Military District (Adelaide) and transferred to the 3rd Military District (Victoria) on 19 June 1916 as a Corporal. He was transferred on 14 July 1916 to 2nd Military District (New South Wales) and started his course at the Engineer Officers Training School at Roseville.

Moffitt embarked again for France, this time aboard HMAT A46 *Clan McGillivray* on 10 May 1917 as a sapper with the February 1917 FCE reinforcements. He was appointed a Sergeant for the voyage only and upon arrival on 28 July was assigned to 3 FCE. He reverted to sapper on 3 August at Brightlingsea; he was immediately appointed Acting Corporal, which he tried for three weeks but reverted to sapper again on 25 August when he proceeded overseas to France and was taken on the strength of 3 FCE on 1 September. On 20 January 1918 he was promoted to Lance Corporal but soon he was on his way to officer school in England. He was appointed 2nd Lieutenant on 1 June 1918 and re-joined his unit on 8 July, whereupon he was transferred to 1 FCE on 11 July. A month later he was wounded in action, with a gunshot wound to the hand. After treatment in England – during which time he was promoted to Lieutenant - he returned to the unit on 10 October.

After the armistice he contracted influenza and after recuperating in hospital re-joined his unit on 6 December. He spent 1919 undertaking non-military employment. He resigned his appointment on 10 March 1920 to return to Australia via America and Canada.

Frank Roland Morris

Lieutenant Frank Roland Morris (41) was a railway engineer from Queensland when he enlisted on 4 July 1916. From 1 November 1916 he attended the Engineer Officer Training School, Roseville, first as Honorary Corporal between 1 and 22 November, then as Acting Corporal between 22 November and 13 December, and Acting Sergeant between 13 December and 21 April 1917, when he was transferred to the Engineer Depot at Moore Park.

He was appointed 2nd Lieutenant on 13 August and embarked aboard HMAT A54 *Runic* in Sydney on 22 March 1918, arriving in London on 24 May; during the voyage he was Ship's Quartermaster as the EOTS had been closed. He took a week's leave and then reported for duty with the AIF in England.

After proceeding overseas via Folkestone he was taken on the strength of 1 Pioneer Battalion in the field on 25 September 1918. He was promoted to Lieutenant on 25 December 1918. He made plans to return to Australia via America to study railway construction methods there but passage was difficult to find.

He was granted non-military employment leave between 9 July and 31 October 1919 but the leave was cut short prior to his embarkation aboard the *Nestor* on 1 November for return to Australia. His appointment was terminated on 9 January 1920.

Ernest Vincent Raymont

Lieutenant Ernest Vincent Raymont (12) who enlisted on 23 October 1915 gave his occupation as 'farmer' and his address as Woody Point, Brisbane, Queensland. He was assigned to C Company of 3 Pioneer Battalion, and when he embarked aboard HMAT A62 *Wandilla* on 6 June 1916 it was with the rank of Sergeant.

He attended Pioneer school at Reading on October 1916 and on completion proceeded overseas to France on 24 November to his unit. He was promoted to 2nd Lieutenant on 11 May 1917 and on 27 June was wounded in action with severe gunshot wounds to his right arm and thigh and left leg and shoulder. He was evacuated to England for treatment, during the course of which he was promoted to Lieutenant. By 17 November he was at Parkhouse no. 3 Camp, where he met Alfred. He returned to France and his unit on 10 March

1918. He suffered a bout of influenza in early July, causing his hospitalisation at 8 General Hospital, then returned to his unit. He took leave in early September in England and was back with his unit by the Armistice.

In 1919 he worked with the Australian Army Pay Corps in London. He returned to Australia on 20 April aboard the *Boonah*. His appointment was terminated on 18 October.

Eric Houghton Rhodes

Lieutenant Eric Houghton Rhodes (11), a structural engineer, enlisted on 16 February 1916. His record of service makes no explicit mention of his attendance at the EOTS Roseville, but his statement of service does refer to the 'EOT School' between 20 November 1916 and 21 April 1917.

The record of service also includes a statutory declaration as to his date of birth, which was signed at Roseville on 17 April 1917, and his application for a commission in the AIF, which refers to 'Officers School Duntroon and 5 months EOTS.' On 21 April he was transferred to the Engineers Depot at Moore Park and given the rank of acting Sergeant. He was appointed 2nd Lieutenant on 21 May and embarked aboard the *Euripides* on 31 October 1917.

On arrival in Devonport on 26 December he marched into Parkhouse no. 3 Camp, and then to the Engineers Training Depot at Brightlingsea on 23 February 1918. He proceeded overseas via Southampton on 3 April 1918 and arrived on 4 April at the AGBD at Rouelles.

On 10 April he marched out to 3 Division Engineers and then, on 17 April, he was taken on the strength of 10 FCE; he transferred to 11 FCE on 25 April. He was promoted to Lieutenant on 23 May. He went on leave to the UK on 8 October and returned on 18 October. He remained with 11 FCE for the rest of the war.

He left for return to Australia on 28 August 1919 aboard the *Kanowna* and his appointment was terminated on 26 November.

George Percy Ross

Lieutenant George Percy Ross (16), a civil engineer, enlisted on 27 June 1916 and was transferred to the Engineer Officers School in Sydney on 1 July 1916 with the rank of Honorary Corporal. He completed that assignment on 9 December 1916 with the rank of acting Sergeant. He was appointed 2nd Lieutenant on 21 May and embarked aboard the *Euripides* on 31 October 1917.

On arrival in Devonport on 26 December he marched into Parkhouse no. 3 Camp, and then to the Engineers Training Depot at Brightlingsea on 23 February 1918. He proceeded overseas via Southampton on 3 April 1918 and arrived on 4 April at the AGBD at Rouelles.

On 10 April he marched out to 1 Army Troop Company Australian Engineers and was taken on strength on 20 April. He was promoted to Lieutenant on 23 May. After a period at the 4 Army Rest Camp beginning on 26 July he returned to his unit on 7 August and remained with it to the end of the war, apart from a period of leave between 23 October and 6 November.

He left to return to Australia on 15 May 1919 aboard the *Orontes*. His appointment was terminated on 28 August.

David George Sinclair

Major David George Sinclair (42) was a surveyor from Queensland who enlisted with his younger brother Donald, also a surveyor, on 18 August 1914. They were given consecutive regimental numbers, Donald 5 and David 6. David was a Corporal (promoted 1 September 1914), Donald a sapper, when they embarked aboard HMAT A2 *Geelong* in Melbourne on 22 September 1914 as part of Headquarters, 3 FCE, bound for Egypt.

David's departure for Gallipoli was delayed by a gunshot wound to the neck in Cairo on 5 May 1915, but Donald left with the unit, and was himself wounded in action on 10 May. David re-joined the unit on 26 May. On 26 July David was appointed 2nd Lieutenant and a day later, Donald was promoted to Sergeant. There might have been a few toasts drunk that night. Sadly, however, Donald was killed in action on 1 August; David's reaction can only be imagined. His war continued, however, but only for a time: on 29 August he went to hospital and was diagnosed with dysentery; he was transferred to London for treatment and recuperation and was declared fit for active service on 14 January 1916. In the meantime he had been promoted to Lieutenant on 1 December.

He re-joined his unit on 13 May 1916 and was promoted to Captain on 21 July. He remained with 3 FCE until 4 January 1918 when he was detached for a period to 3 Infantry Brigade Headquarters, returning on 16 January; he was again detached, this time to Royal Engineers School, Blendeques between 8 February and 2 March. On 4 April he was promoted to Temporary Major, on 7 April he was mentioned in despatches, and on 19 April he was taken on the strength of 1 FCE as Officer Commanding. His promotion to Major was confirmed on 30 July and dated from 4 April. On 16 July he left for 14 days furlough in England during which time he was admitted to hospital, diagnosis not recorded. It was not until 23 August that he re-joined the unit, and on 29 September he left for furlough in Australia on special leave for those who had joined up in 1914.

He embarked aboard SS *Olympic* on 3 October to travel via America. His appointment was terminated on 16 January 1919.

John Edward Graham Stevenson

Captain John Edward Graham Stevenson (22) was a surveyor before enlisting on 14 December 1914, when he was assigned to the 5th reinforcements, 3 FCE.

He embarked as a sapper aboard HMAT A55 *Kyarra* on 16 April 1915 and was taken on the strength of his unit at Gallipoli on 22 June. After being promoted to 2nd Lieutenant on 22 August, he transferred to 2 FCE on 25 August and was promoted to Lieutenant on 1 December.

After the withdrawal from Gallipoli later that month Stevenson went with his unit to France where, during the operations *at Pozieres between 20 and 25 July he had charge of a section building strong points in captured trenches and displayed great coolness and ability under heavy shell fire, setting a splendid example to his men.*[42]

On 13 August 1916 he transferred to 1 FCE and shortly after, on 16 August, he was detached for duty at Brigade Headquarters. The citation for the award of the Military Cross continues:

> *Again during the operations from 15 to 22 August he acted as second in command of 1 FCE and was attached to Brigade Headquarters and did very valuable work in that capacity.*

The award did not come through. After a furlough in England he re-joined his unit on 6 October. He was promoted to Captain on 1 January 1917 and served as acting Officer Commanding the Company between mid-January and mid-March. He remained with the unit in France until 28 September 1917 when he was seconded for duty with the Engineers Training Depot at Brightlingsea, England.

On the afternoon of 21 September and the morning of 22 September he displayed great courage and tenacity in carrying out his role of reconnaissance officer, keeping his company headquarters supplied with valuable information regarding proposed engineer work and accommodation and shelter for the troops; for this action he was awarded the Military Cross on 1 January 1918.[43] The plan was for him to return to his unit early in 1918, and he got as far as Rouelles when he was attached to, and later taken on the strength of, 1 Division Engineers Headquarters as Adjutant. He relinquished this appointment and re-joined 1 FCE on 24 April 1918. After a brief attachment to 2 FCE he spent September 1918 at 4 Army Infantry School, re-joining his unit on 1 October and enjoying a period of leave in the UK between 4 and 21 October.

On 28 March 1919 he embarked aboard *City of Poona* for return to Australia. His appointment was terminated on 14 September 1919.

Alexander Jarvie Hood Stobo

Lieutenant Alexander Jarvie Hood Stobo (41) was from 1 Battalion AIF but was attached 119 British Bde France at the time he signed the book. His signature was 'Alister Stobo,' a family name. Born on Harwood Island in the Clarence River, NSW, he was a medical student at the University of Sydney when he enlisted on 24 September 1915, having passed his first-year examinations.

He was assigned as a private to 12 reinforcements, 4 Battalion, and embarked aboard HMAT A7 *Medic* on 30 December 1915 from Sydney. He was taken on the strength of 1 Battalion in Egypt on 17 March 1916 and travelled with the battalion as it moved to France.

He was promoted to Lance Corporal on 6 May and wounded in action – a gunshot wound to the scalp, so a close-run thing – on 28 June.

He re-joined the unit on 6 July but was gassed on 25 July; discharged to duty on 30 July he reported to hospital again on 4 August with a hernia. After treatment he re-joined the unit on 26 August. On 25 September he found himself on command at Divisional School of Instruction 'in the field', which was on the front line as 1 Battalion relieved 6 Battalion that day at Ypres Asylum. He returned on 13 October, only to be assigned on 1 November to the Officers' School of Instruction at Balliol College, Oxford as a member of 6 Officer Cadet Battalion.

On 1 March 1917 he was appointed 2nd Lieutenant and later that month he returned to France. Wounded in action again on 7 May he re-joined 1 Battalion on 28 May, and was sent to sniping school from 10 to 24 June; on 20 July he was attached to Brigade Headquarters for intelligence duties and from there he went to Musketry Instructor's Course at Small Arms School, returning eventually to his unit on 7 December, having been promoted to Lieutenant on 24 November. He returned to 1 Battalion and then was absent for two weeks' leave in the UK at the end of January 1918, Corps Intelligence School for a fortnight from 28 June, an attachment to 119 Brigade (see above) and leave in the UK from 8 to 26 August; later he spent time with 1 Brigade AIF Headquarters, 30 Division of the US Armies, and secondment as Intelligence Officer with 1 Brigade AIF HQ, and that was just September.

He embarked on 16 March 1919 for return to Australia initially aboard the *Czaritza* but was transhipped to the *Dunluce Castle* at Alexandria on 7 April and to *City of Poona* in Sydney on 18 May, disembarking at Brisbane on 20 May. His appointment was terminated on 12 July. He became a doctor in Sydney.

George Lancelot Allnutt Thirkell

Captain George Lancelot Allnutt Thirkell (34) was an engineering graduate of the University of Tasmania, working as a draughtsman and Clerk to the Engineer of Railway Construction in Tasmania, when he enlisted on 26 August 1914 in 3 FCE.

He embarked as 2nd Lieutenant aboard HMAT A2 *Geelong* on 22 September 1914 for Egypt, landing eventually at Gallipoli. However, he developed a severe case of enteric fever there and returned to Australia, disembarking at Melbourne on 1 October 1915.

Whilst his activities during 1916 are not detailed on his record of service, he was photographed in June 1916 in a group portrait of C Squadron, Engineer Officers School of Instruction at Moore Park, Sydney. He is sitting in the first seated row, the row usually occupied by the instructors in the school; on his left is Lieutenant Frederick Henry Morton, whose record shows that he was at the school between 4 December 1915 and 4 September 1916, six months of which were as an instructor.

On 11 November Thirkell embarked aboard HMAT A29 *Suevic* to return to the Western Front; there was a slight delay after he arrived in England for him to recover from a bout of pneumonia, but on 23 April 1917 he was taken on the strength of 3 FCE. A few days later, on 11 May, he marched into 17 FCE, but returned to 3 FCE on 7 November, when he was promoted to Captain. He remained with 3 FCE for the rest of the war.

He took a period of leave in the first half of 1919 and non-military employment in the second and embarked on 22 January 1920 for return to Australia aboard HT *Friedrichsruh*. His appointment was terminated on 10 May 1920.

Group portrait of C Squadron, Engineer Officers School of Instruction, Moore Park, June 1916. George Lancelot Allnutt Thirkell is seated in the middle of the second row.

Howard George Tolley

Major Howard George Tolley (18) was a licensed surveyor and irrigation engineer when he enlisted on 26 April 1915. He was assigned to 9 reinforcements, 1 FCE on 16 June at the Engineer Depot in Moore Park, Sydney. On 3 September he qualified at a competitive examination for his first appointment as a 2nd Lieutenant and was appointed to that rank on 24 September.

He embarked aboard HMAT A23 *Suffolk* on 30 November in Sydney for Egypt with 7 FCE, where on 4 March 1916 he was promoted to Lieutenant and, the next day, transferred to 12 FCE on its formation; that same day, after having spent only one day as a Lieutenant, he was promoted to Captain.

On 4 June he embarked aboard HMT *Scotian* for Marseilles and the Western Front, arriving at La Clytte on 23 June. On 21 March 1917 he was promoted to Major and joined 4 FCE as Officer Commanding. On 1 June he was mentioned in despatches

...for general good work and devotion to duty at Pozieres, Ypres and Gueulecourt, August 1916 to February 1917. This officer's professional knowledge is of a very high order, and has been of great value on many occasions, particularly as regards water supply. His personal conduct has been very gallant and his care for his men most marked. He has done both good engineering work, and good work when acting as a company commander.[44]

On 19 November he was awarded the Distinguished Service Order for conspicuous technical skill, gallantry, and devotion to duty during operations in front of Zonnebeke on 24, 25 and 26 September. The citation reads –

He personally reconnoitred for forward routes for the advancing Infantry and for Divisional Pack Trains, having to go through the barrage put down by the enemy to do this. During his reconnaissance, whilst the attack was in progress, he explored all the recently captured enemy concrete dug-outs in the Brigade area, assisted to supervise the work of the Infantry in digging in on the lines of the 1st and 2nd objectives, and reconnoitred and reported on the condition of the forward roads. His report was delivered personally at Divisional Headquarters on the evening of the 26th. It was due to his energy and personal supervision that the getting up of Engineer stores, the laying out and lighting of communication routes, was so successfully carried out in spite of the fact that there were only three days available for preparatory work. He has at all times set a fine example of courage and cheerfulness under the most difficult conditions, and his high technical qualifications have always been of the greatest value.[45]

After a period of leave in the UK in early September 1918 he spent the period between 24 September and 2 October attached as liaison officer to 102 Regiment United States Engineers.

During the whole period he was attached he did most meritorious service for the above unit and his efficiency was very highly commended by the Corps Engineer of the II American Corps.

On 29 September ... *he went forward with a party of American Engineer Officers and runners to establish a Forward Report Centre. The location previously selected was found to be in hostile occupation. The shelling was heavy and despite the fact that the whole of this party had become casualties, Major Tolley single-handed and unarmed still pressed on. He came upon a party of the enemy and succeeded in capturing seven prisoners by threatening them with a German Stick Bomb which he picked up nearby. Later in the day he organised defense [sic, it must be the American influence] against Enemy Aircraft, for which no provision had been made and in this way was directly responsible in bringing down at least one Enemy Aeroplane which was causing many casualties with its machine guns.*

For this action he was recommended for the American Distinguished Service Medal.[46]

On 16 February 1919 he was temporarily appointed Commander Royal Engineers of 5 Australian Division, a position he relinquished on 3 May to undertake non-military employment between 7 May and 7 July to undertake professional visits.

Lieutenant Tolley (below) embarked for return to Australia aboard the *City of Exeter* on 12 July and his appointment was terminated on 18 October. He went on to serve in World War II.

Walter Howard Tunbridge

Colonel Walter Howard Tunbridge CB CMG CBE VD (36), Senior Mechanical Transport Officer and Director of Mechanical Transport Services, an architect by profession, began his military work in 1889 with his appointment as acting Lieutenant in the Queensland Defence Force. He was appointed Lieutenant in the Townsville Mounted Infantry on 1 January 1890 and Captain on 24 July 1892. He passed the examination for Major in 1898 and was appointed to that rank on 14 January 1899. He was appointed as a special service officer with the Queensland contingent for the South African war and commanded the third Queensland contingent of troops that left on 1 March 1900. After action at the relief of Mafeking and the defence of Eland's River, he was appointed commander of the second Australian regiment. He was mentioned in Lord Roberts's despatch for meritorious services and made a Companion of the Bath (CB) in April 1901. He returned to Queensland and led the welcoming procession through Brisbane streets on 13 June.

He was appointed aide-de-camp to the Governor-General in August 1902 for a period of 5 years, and again in 1907 for a period of two years, on the basis of gallantry and distinguished service in the field. He was made a brevet Lieutenant-Colonel on 28 November 1902.

After a move to live in Melbourne, a career in business and a period in the Reserve of Officers of the Australian Military Forces, he resumed his military career in the Divisional Ammunition Park. By April 1915 he was in England with the DAP. In 1916 his gallant and distinguished conduct in the field was drawn to His Majesty's notice by General Haig. In April 1917 he was appointed Senior Mechanical Transport Officer and Director of Mechanical Transport Services for the AIF. On 24 September 1917 he was made Brevet Colonel, on 1 June 1918, Colonel, on 1 January 1920, Honorary Brigadier General. On 1 January 1918 he was made CMG, and on 29 May CBE. On 10 April 1918 his retirement was postponed until three months after the war. His appointment was cancelled on 28 June 1920.

Frank Meredith Turton

2nd Lieutenant Frank Meredith Turton (15) was a clerk when he enlisted on 2 December 1915 and was assigned to A Company, 44 Battalion, AIF as a private.

He embarked aboard HMAT A29 *Suevic* in Fremantle on 6 June 1916, arriving in England on 21 July. The battalion transferred to Lark Hill Camp and commenced field training on 22 July. After a brief brush with influenza Frank proceeded overseas to France with 44 Battalion on 25 November.

He was promoted to Lance Corporal on 27 January 1917, when the battalion was under heavy enemy fire in the trenches at Armentieres. He was detached to attend 3rd Division School on 3 April and 2 Anzac Corps School on 7 April. This was a man on the way up. On 5 May he reported for duty with the no. 2 Officer Cadet Battalion at Pembroke College, Cambridge, for officer training.

He was appointed 2nd Lieutenant on 1 September 1917 and posted to General Infantry reinforcements to await assignment to a battalion. In the event he was admitted to 3 London General Hospital on 10 September and discharged to Sutton Veny on 13 November and Parkhouse no. 3 Camp on 17 November. There he remained, and met Alfred, until 1 April 1918 when he was transferred from General Infantry reinforcements to the Permanent Supernumerary List and posted for duty with AIF Depots in the UK. On 14 July he was promoted to Lieutenant and by 15 July he was at no. 2 Command Depot Weymouth.

He was admitted to 3 London General Hospital on 7 February 1919 for an undetermined complaint. He embarked on 16 March for return to Australia initially aboard the *Czaritza* but was transhipped to the *Dunluce Castle* at Alexandria on 7 April. His appointment was terminated on 3 July.

Edward Charles White

Lieutenant Edward Charles White (25), a builder and architect, started his enlistment process on 21 June 1916 and finalised it on 22 August 1916. He completed his attestation paper at the Roseville EOTS, which he attended from then until 1 February 1917, when he was transferred to the Engineers Depot Sydney with the rank of acting Sergeant. He was appointed 2nd Lieutenant on 21 May and embarked aboard the *Euripides* on 31 October 1917.

On arrival in Devonport on 26 December he marched into Parkhouse no. 3 Camp, and then straight to hospital. Recovered by 3 January 1918, he returned to Parkhouse and marched into the Engineers Training Depot at Brightlingsea on 17 February.

He proceeded overseas via Southampton on 3 April 1918 and arrived on 4 April at the AGBD at Rouelles. On 10 April he marched out to 2 Division Engineers and then, on 18 April, he was taken on the strength of 6 FCE; he transferred to 2 FCE on 10 May. He was promoted to Lieutenant on 23 May. He remained with 2 FCE for the rest of the war, apart from a period of leave in the UK between 9 and 26 September.

He left for return to Australia aboard the *Ormonde* on 16 June 1919. His appointment was terminated on 27 August.

13

The State Aviation School at Richmond

The NSW Premier, Mr Holman, established a NSW State Aviation School when it became clear (to him, at least) that the entrants selected for entry to the Point Cook flying school in Victoria did not have as high a representation from NSW as Holman thought appropriate, given the history and size of his state.

The purpose of the school was 'to train men, who are qualified for the Army, in every branch of the art of aviation prior to their military training.'[47]

The school was a great success, largely due to the appointment of Lieutenant William Stutt (formerly of the RFC) as Chief Instructor and the generous provision of funding for the trainees' pay and flying time. Of the first intake of 20 students, 19 gained their pilot's certificate; of these, 10 were admitted to the Australian Flying Corps, one was retained by the school as an assistant instructor, and the remainder were to be sent by the state government to England for absorption into the Imperial Air Service.[48] Maybe Holman was right.

The second and third intakes were smaller. This allowed more flying time for each student, which shortened the duration of the course. The government encountered some turbulence in the administration of the school, and after an initial period under a committee of control the School was placed under the control of the Minister for Education as part of the Technical Education Department.[49]

The signers who indicated that they were from the RFC were drawn from the second and third intakes into the school.

THE SECOND INTAKE: Back row (left to right, * indicates signer): S C Francis*, A C Le Grice*[50], D R Williams, W J Stutt, R H Chester (staff), F. C. Collins, L. C. Royle*. Middle row: J H Summers, D Hudson, B Lucy*, B B Sampson*, W R Boulton, M A Watts*. Front row: H G Murray, L Audet, G V Oxenham, W A McDougall.[51]

THE THIRD INTAKE: Back row (left to right, * indicates signer): A S Edwards*, A B Bargwanna*, R H Chester, W J Stutt, H G Murray (staff), G Dobell*, D W Cullam, Cyril Charles Worboys*. Front row: K Mathieson; J H Crews, R S Nickoll*, G V Wicks, L G Cole*.

14

Picture Credits

Alfred's autograph book - *Family collection, C Edye scan*
Alfred Richmond Wright - *Family collection, C Edye scan*
Alfred's vest pocket camera - *Family collection, Ian Thom photograph*
Honest toilers and Smoko sport - *Family collection, Ian Thom scan*
2nd Lieutenant Wright with his kit - *Family collection, Ian Thom scan*
Soldiers' mess (2) - *Family collection, Ian Thom scan*
EOTS No. 1 Section - *Family collection, C Edye scan*
Bridge over Moores creek at Roseville - *Family collection, Ian Thom scan*
EOTS June 1917 Reinforcements - *Family collection, Ian Thom scan*
Field Engineers 1917 Unit flag - *Family collection, Ian Thom photo*
HMAT A14 *Euripides* - *Hamel family collection, C Edye edit*
The bridge and the men who built it - *Family collection, Ian Thom scan*
Garnet Halloran - *University of Sydney Archives G3 224 0684*
Lieutenant Cyril Lawrence - *AWM3794854, C Edye edit*
Photographs of Harbonnieres and Villers-Bretonneux - *ETT Imprint*
Where George Kerslake was killed - *AWM P03483.027, C Edye edit*
Wreck of the *Iron Chief* - *ETT Imprint*
William Edmund Fitzgerald - *AWM P09650.001, C Edye edit*
Lieutenant William Montgomerie Fleming -
From https://en.wikipedia.org/wiki/C Edye edit
Colonel Tunbridge, Captain Maplestone - *AWM E04287, C Edye edit*
Group Portrait C Squadron - *AWM P09176.002*
Lieutenant Howard George Tolley - *AWM H00129, C Edye edit*
Aviation School second intake - *AWM P00731.005, C Edye edit*
Aviation School third intake - *AWM P00731.006, C Edye edit*
And a little bit of fun to end it all - *AWM A02853, C Edye edit*

15

And a Little Bit of Fun at the End

This is a group portrait photograph of officers of 17 Battalion taken in Egypt on 15 August 1915 (AWM A02853). It was taken too early for inclusion in this document but qualifies anyway because it shows Captain Llewellyn Griffiths (third from the left in the front row) (17) and Lieutenant (later Major, MC) Herbert Leslie Bruce (seventh from the left in the middle row), the author's great-uncle

16

Endnotes

1 *Northern Star*, Saturday 19 March 1932, page 6
2 Current circumstances mean that further investigation of this information, which would need to be done in person at NSW State Archives, is not possible.
3 In the same issue of the Commonwealth of Australia Gazette (3 February 1916, p 224), the secondment of E V Stratton from the Senior Cadets whilst serving in the Australian Imperial Force was published. Stratton appears later in Alfred's story.
4 The spelling is impressionistic
5 The numeral in the brackets is the number of the page opening in the autograph book.
6 Bold type indicates that there is a brief biography of the person concerned in the section Extended biographical notes.
7 Soldier, of Victoria
8 Commonwealth of Australia Gazette, 8 February 1917, page 246
9 National Archives of Australia: B2455, WRIGHT A R, p 16
10 Stanton-Cook's biography appears in Vol III of *Rallying the Troops*, page 342
11 Royle's biography appears in volume IV of *Rallying the Troops*, page 585
12 According to their respective records of service, both Ferrier and Alfred Wright were at the AIBD between 3 and 10 April. With the evidence available it is not possible to reconcile the Autograph book signature date (which is clearly sometime in March) with the two official records.
13 Soldier, of Elsternwick, Victoria
14 Engineer, of Chatswood, NSW
15 Blacksmith, of Durham, England
16 Survey draughtsman, of Tempe NSW
17 Architect, of Neutral Bay, NSW
18 1 FCE war diary for 1 May 1918 at https://www.awm.gov.au/collection/C1347283 accessed 20200330

19 1 FCE war diary for May 1918, op. cit. 'Wiring' refers to barbed wire.
20 Engineer and licensed surveyor, of Mt Lawley, WA
21 1 FCE war diary for 13 June 1918 at https://www.awm.gov.au/collection/C1355036 accessed 20200330
22 National Archives of Australia: B2455, WRIGHT A R, p 28
23 1 FCE war diary for 3 August 1918 at https://www.awm.gov.au/collection/C1347285 accessed 20200330
24 14 August according to the war diary, 15 August according to Fitzgerald's record of service.
25 Details from Kerslake's record of service, at NAA: B2455, KERSLAKE GEORGE KNIGHT accessed 20191122
26 See, for example, *The West Australian*, Saturday 16 November 1918, p1
27 Age difference calculated from a letter from Audrey published in *The Daily News*, Saturday 24 Aug 1907, page 15
28 Engineer, of Armadale, Victoria
29 Soldier, of Perth, WA
30 Material in this paragraph is drawn from reports in *The Daily News* of Tuesday 2 March 1920, pages 6 and 7
31 *The Argus*, Monday 8 March 1920, page 6
32 *Northern Star*, Tuesday 30 November 1920, page 5. Family research shows that Charlotte's brothers 2221 Alfred Arthur and 4833 William Vincent Valentin had already enlisted (on 6 and 7 May 1915 respectively) in the AIF "to fight the Germans"
33 The 1930 and 1931 Electoral Rolls have Alfred living at the Ordnance Building.
34 *Northern Star*, Saturday 19 March 1932, page 6
35 *The Sydney Morning Herald*, 19 March 1932, page 11
36 Broughton, Francis William Walford record of service at NAA: B2455, BROUGHTON F W W accessed 20200204
37 *Daily Examiner*, 28 April 1917 p 4
38 The citation for the award is at https://www.awm.gov.au/collection/R1595138 accessed 20200325
39 Details of wounding from 5th Australian Ammunition Sub-Park war diary at https://www.awm.gov.au/collection/C1340607?image=10 accessed 20200209
40 The citation is at https://www.awm.gov.au/collection/R1629331 accessed 20200209
41 Details of career from NAA: B2455 MAPLESTONE Lyndsay Lorrance (sic) accessed 20200209
42 The citation is at https://www.awm.gov.au/collection/P10223487 accessed 20200325

43 The citation is at https://www.awm.gov.au/collection/R1627436 accessed 20200325
44 The citation is at https://www.awm.gov.au/collection/R1624044 accessed 20200325
45 The citation is at https://www.awm.gov.au/collection/R1586602 accessed 20200325
46 The citation is at https://www.awm.gov.au/collection/R1586924 accessed 20200325
47 *Referee*, 12 September 1917, page 13
48 *The Sydney Morning Herald*, 17 January 1917, page 7
49 *Sunday Times*, 15 July 1917, page 8
50 Le Grice was Holman's driver
51 Biographical pieces for Lucy, Royle and Summers can be found in *Rallying the troops*, A World War I commemoration by Ku-ring-gai Historical Society

17
Index

*Autograph Book Signers

Ainslie*, Edgar 39
Allan*, James W 38
Allpress 44
Anchises 49
Anderson*, Alick T 20
Argyllshire 66
A R Wright & Company 43, 44
Ascanius 61, 63
Bailey, Herbert A 19
Bailey*, Herbert F 20
Ballina 10
Beltana 53
Boake*, Samuel 21, 23
Boonah 65
Brightlingsea 25, 33, 48, 51, 53, 54, 55, 57, 59, 63, 65-66, 68, 75
Broughton*, Francis W 23, 49, 81
Brown*, Allan I 39
Brown*, Allan M 30, 35, 49
Bruche, Julius H 36, 39
Burge, Edward 42
Butler*, Ethel B 37
Buttner*, Adolph R W
Canberra 55
Carr*, Gerald M 25, 50
Carroll*, A 37
Christie*, David 24, 51
Christies Auctioneers 48
City of Execter 72
City of Poona 68, 69
Clan McGillivary 63
Clarke*, Reginald E 38
Clay*, William R 24
Cobcroft*, John V 21
Cole*, Lew G 22, 77
Cook*, Norman 16, 20, 22, 51
Cooper*, Cyril M 21
Cowled*, Agustus J 21
Czaritza 55, 69, 74

Darby*, Harold F 37
Darlow*, George P 21
Delta 59
Demobilisation Department 36, 39
Dobell*, G 23, 77
Donaldson*, James B 40
Dunluce Castle 69, 74
Duntroon Officers School 60, 65
Durrant*, James M A 38, 39
Edwards*, Alan S 24, 77
Engineer Officers Training School (Moore Park) 11, 55, 64, 65, 70, 71
Engineer Officers Training School (Roseville) *8, 11, 13, 14, 15, 16, 28, 28, 51, 55, 57, 63, 64, 65, 75, 78*
Euripides 6, 16, 17, 18, 19, 22, 24, 25, 29, 51, 57, 65, 66, 75
Ferrier*, Lester 26, 52, 80
Fifield*, Frederick W 21
Finlay*, James H 21
Fitzgerald*, William E 32, 33, 34, 53, 78, 81
Fleming*, William M 22, 54, 78
Fletcher*, Cecil J 38
Francis*, Sid C 21, 77
Fraser*, Alexander 39
F R Strange Auctioneers 48
Friedrichsruh 37, 38, 41, 42, 50, 57, 60, 70
Gannon*, Josephine M 37, 55
Gaskell*, George E 39
Geelong 66, 69
Gill*, D H A 40
Gossip*, James 22
Greenwood*, J C 30, 35, 49
Griffiths*, Llewellyn 39, 79
Halloran*, Garnet 22, 27, 78
Harris*, William D 22
Heath*, Henry F T 39
Hetherington*, Henry F 27, 29, 32, 33, 55
Higgins*, Walter H 33, 56, 59

Hohnen*, Frederick H 21
Holman, William 76, 82
Hope*, Alfred J 22
Hopton*, Charles A 40
Hororata 59
Houghton*, William S 39
Hubbard*, Frank J 40
Hunsted*, Henry 33
Hunt*, Marion 38
Iggledon*, S E 22
Ionian 58
Iron Chief 45, 46, 78
Jenkins*, Charles A 24
Jennings*, Charles J 38, 57
June 1917 Reinforcements 16, 21, 78
Kanowna 65
Karoola 63
Kerr*, Robert B 21
Kerslake*, George K 16, 21, 27, 28, 29, 30, 32, 33, 34, 57, 78, 81
Koch*, George H 25, 58
Kyarra 66
Lavelle, Digby and Eva 44
Lawrence*, Arthur P 39
Lawrence*, Cyril (Oscar) 26, 27, 32, 56, 59, 78
Lawson*, John 38
Le Grice*, A C 21, 77, 82
Le Sapper*, A 24
Lecky*, Joseph 23
Lee*, Hector 37
Lindsay, F A 44
Lismore 10, 47
Loughnan*, John 39
Lucy*, Brian 23, 77, 82
Maclean*, Jonathan M 21
Maplestone*, Lyndsay T 38, 60, 61, 78, 81
McAnna*, James N 37
McArthur*, Alexander D 25, 61
McConnel*, Samuel 33, 62
McNicol*, Daniel A C 38
Medic 68
Melville*, E G 20
Mendelsohn*, Harris 38
Militades 55

Moffitt*, Frank G 30, 32, 63
Moleghein Farm 30
Mollemodge*, M A 41
Monash, Sir John 36, 38
Moolton 55
Moore*, Donald T 39
Moran*, John T 23
Morris*, Frank R 37, 64
Mt St Quentin 33
Murray*, Arthur L 39
Murray, H G 77
Nestor 61, 64
Newalia 33
Newton*, Alfred J 24
Nickoll*, Roger 23, 77
Norman 52
Olympic 66
Ordnance Building 43, 44, 45, 81 *Orita* 54
Ormonde 55, 75
Orontes 66
Orr*, Thomas 39
Osmaniah 55
Osterley 56
Palmer, Audrey V 34
Palmer*, Charles R 38, 40
Parkhouse 25, 49, 50, 51, 55, 57, 58, 61, 64, 65, 66, 74, 75
Pateman*, T 39
Paterson*, S E 22
Penrose*, William G 22
Persic 60
Peterson, Lieutenant 29, 30, 32, 33
Pinkstone*, Sidney A 24
Plunkett*, Gunning F 39
Ponton*, J 41
Poole*, Arthur L, 39
Poole, Louis W 28
Public Trustee 48
Pugh*, F W 25
Quigley*, Gavin M 39
Raymont*, Ernest V 25, 64
Rhodes*, Edward H 16, 21, 22, 65
Roberts*, Stephen R H 24
Root*, A N W 39
Ross*, George P 16, 21, 48, 66

Rouelles 26, 51, 57, 65, 66, 68, 75
Royle*, Raynes L C 24, 77, 80, 82
Runic 64
Sampson*, Burton B 23, 77
Sanderson*, William L, 37
Saunderson*, William, 22
Savage*, James E, 40
Scotian 71
Shannon*, William F, 40
Sinclair*, David G, 27, 30, 34, 66
Smith*, George H, 20, 21
Stanton-Cook*, Lance H, 23, 80
Star of Victoria 62
Steele*, C S, 35
Stewart Smith* J C, 24
Stobo*, Alexander J H, 30, 68
Stone*, H Kenunga, 20
Stone*, Gladys I, 20
Stopford*, Grosvenor F, 22
Stratton*, Erith V, 40, 80
Strazeele, 28, 30
Suevic, 70, 74
Suffolk, 71
Taylor*, W H, 22
Thirkell*, George L A, 38, 69, 70
Thomas*, William J, 23
Thorburn*, T, 24
Tincourt Military Cemetery, 33
Tolley*, Howard G, 37, 71, 72, 78
Tunbridge*, W H, 40, 61, 73, 78
Turton*, Frank M, 25, 74
Ulysses, 52
Valentin, Alfred A, 81
Valentin, Charlotte A, 10
Valentin, Wilfred V, 81
Vaux-sur-Somme, 32
Viccars*, Stanley G, 22
Villers-Bretonneux, 31, 32, 78
Walker*, Harrild M, 22
Walklate*, Charles G, 22
Walsh*, Clement R, 40
Wandilla, 64
Watson*, Ian T, 40
Watts*, M A, 24, 77
Wauhope*, Edward, 23

White*, Edward C, 16, 22, 29, 75
Williamson*, Keith, 24
Worboys*, Cyril C, 24, 77
Wright, Alfred (Father), 10
Wright, Alfred R, 6, 7, 8, 10, 11, 13, 15, 16, 20, 21, 25, 26, 27, 28, 30, 32, 33, 34, 35, 36, 37, 38, 40, 41, 42, 43, 44, 45, 46, 47, 48, 50, 52, 55, 58, 64, 74, 78, 80, 81,
Wright, Clara née Carr (Mother), 10
Wright, Daphne, 10, 47
Wright, Rex, 10
Wright, Valerie, 10
Ypres, 56, 59, 69, 71
Zealandia, 62

www.ingramcontent.com/pod-product-compliance
Lightning Source LLC
LaVergne TN
LVHW091552070426
835507LV00010B/807